The Theory Of Despace

Reelin Show

First Edited **Aug-2009**

Revised **Aug-2010**

TAIWAN

Preface

We, human beings are equipped with all what we need to sense anything in our surroundings since we've been born. We praise for how we can sense and feel, but we are also blinded by it. We are all imprisoned by our sensitivity and not even realized. However, should all what we've believed so far always be what we've believed so far? Don't we deserve a chance to think once again from the start?

I think the nature of the universe is the simplest one among anything we knew. And this simplest nature of the universe forms our complex world. However, the conclusion from my repeatedly contemplation totally exceeds what I have expected. The nature of the universe is so simple that no other existence in the world can be simpler than it. Furthermore, the greater and the more surprising the theory is, the simpler the theory is normally. So the ideal of this theory is to let everyone to be able to understand it easily and to realize that how surprising it is after their understanding of how simple it is.

These thoughts and theories stated in this paper may still be crude, may not be flawless. However, how abstruse the theory for the universe is, it's just similar to how vast the universe it is. It's nothing but wasting time of trying to totally comprehend it by one's own. Therefore, I hereby disclose my superficial theory to the public. Hope that it can provide something useful to other researchers who are interested in it.

Content

Chapter 1   Space & Despace

Chapter 2   Photon & Halfton

Chapter 3   Ether & Space

Chapter 4   Discussions

Abstract

"For the bubbles in the water, there is no water". This is the basic concept of the Theory of Despace. The Theory of Despace states that the universe is filled up with "space", and matter is the place in the universe without space.

Based upon this postulation, the Theory of Despace created a new model of the basic particles to explain the forces in distance such as Gravitation, electric field and magnetic field. Based upon the model of the electric field in this theory, the Theory of Despace also created a new model of photon which composed of two of the particles — one halfton and one anti-halfton — to explain the wave-particle duality of light. Meanwhile, since the photon is composed of two particles which move in a course of a waveform. For the light speed with respect to the space, it's considered as a wave to be limited to the light speed in that medium. For the drag effect of the space to the photon, it's considered as a particle to become very small.

The Theory of Despace also states that the planetary motion is caused by the space vortex. All the planet systems are free vortices. The speed of the space flow at the outer ring of the vortex is slower than which at the inner ring. When light beam comes from far away and reaches Earth, the vortex speed will vary from 0 to about the revolution speed of Earth around the sun. The light speed with respect to the space will remain constant as light is a wave in the space as its medium, and the direction of the light beam with respect to its source will also remain the same because the drag effect of the space to the photon will become 0 since the photon is a particle and the speed difference of the space flow is extremely small when the photon moves in the space. This solved the paradox between the phenomenon of stellar aberration and the MMX.

CHAPTER 1

SPACE & DESPACE

1.1 Nothing or Something, Infinite or Finite

For the human basic concept of the space and matter, human beings believe that the space is empty and matter is tangible. However, such an acknowledgement does not have any stand. Hereby, we have to ask ourselves the following questions:

1) Is there a beginning of the universe?
2) Is infinity true?
3) Is empty space true?

1.1.1 The Model of the Universe with Empty Space

First, we assume that the space is empty as our traditional concept, then the universe can be shown as the below model:

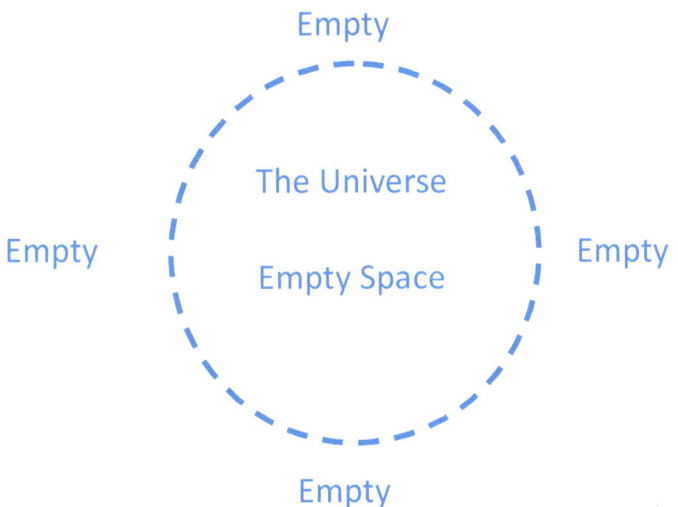

(Fig1.1-1: The model of the universe with empty space)

According to this model, we discern that:

a) The space inside the universe is empty.
b) The place outside of the universe is empty as well.

In this case, we can conclude that if the space is empty, then our universe shall be infinitely large without boundary.

1.1.2 The Model of the Universe with Space Not Empty

However, according to some astronomical observations and theories, our universe might be about 14 billion years old. In other words, there should be a beginning of our world. This means before that point of time, the universe did not exist. And after that point of time, there was the universe. The size of the universe is zero before it exists and might be infinitely large now. Thus, the infinite universe must be changed from a finite one.

However, a finite one could never change into infinity. It's like counting a number. No matter how we count, it's always a number, it will not become into an infinite one. Thus, the universe is a finite one with boundary. The infinite universe is false. And our universe can be shown as the below model:

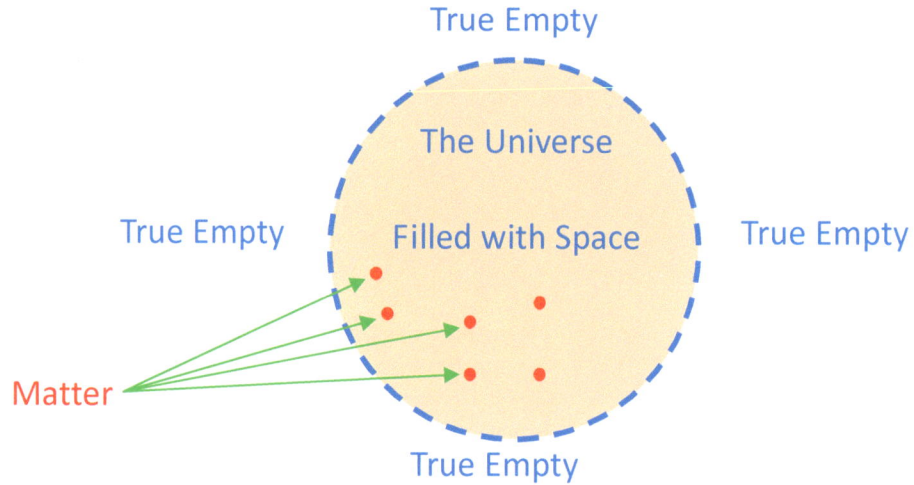

(Fig1.1-2: The model of the universe filled with the space)

Since the universe is not an infinite one, thus the universe should have a boundary. And the place outside of the boundary of the universe is to be defined as a place with nothing, a place (or not even a place) called as *Nihil*, or the true empty. No space, no dimension, no mass, nothing, not even has the property of existence.

Since the universe has a boundary and outside of its boundary is the true empty, thus, the space inside the universe is not empty. The space is something rather than nothing.

On the other hand, matter is contained within the space, and matter can move inside the space. In this case, the space should not be the solid but something like the fluid flowing in dimensions. And hence we can conclude as following:

1) *The universe has its limit and a boundary.*
2) *The space inside the universe is something rather than nothing.*
3) *The space is not solid but something like the fluid.*

1.1.3 Nihil

Nihil, or some others may call it as the true vacuum or the true empty, is the greatest existence in the world which does not have the property of existence, which does not actually exist, however it exists everywhere.

Vacuum is a place full of the space where the air pressure is 0. Nihil is not even a place, not a thing, no space, no dimension. It can be infinitely small or infinitely large since it does not have dimension. It exists in all dimensions though it does not actually exist.

The difference between Zero and Nihil can be clarified as below:

Zero — There is no apple.
There is no apple someplace. The number of the apple is 0. It's countable. It can be measured.

Nihil — There is no apple in the world.
There is no apple such thing in the world. The apple is meaningless. You can't even count it, measure it. However, you still can imagine it to be as whatever you like.

It's great because it does not actually exist, however we use it for everyday. For example: There is Heaven or not. There is Hell or not. They do not actually exist, however they exist everywhere.

1.2 Space and Despace

As we all know that there is space everywhere. Even within the matter with the highest density, the atoms and the particles that form the matter only occupy a small part volume of the space where they are. However, do the atoms and the particles need space? Is there any space where the atoms and particles exist?

I hereby assume that the particles which form the atom do not need the space. More exactly, the particles do not contain the space in themselves. In this case, the place being occupied by the particles will become as holes without the space in the space. On the other hand, no matter what the particles actually are, they can be just taken as the holes without the space in the space. Therefore, we can have our basic definitions as following.

1.2.1 Definition

According to above, we can redefine the universe and the formation of matter. I hereby simply introduce 2 things（there is actually only one thing in the world）:
1) *Space – The one which fills up the world.*
2) *Despace – A hole without the space in the space.*

It seems unnecessary to explain the existence of the space and matter. However, what is space? And what is matter? Some said that the space is a vast of empty vacuum or the *Nihil*. And matter is the substance with shape and mass. This is just a traditional idea without any stand. Sometimes we can easily be deceived by what we see and how we feel. Matter is right there. It's so tangible, it is so concrete. Is that true? Actually, what we can see, what we can hear and even how we feel, they are series of complex processes and combinations of forces and reactions. The human bodies or the measuring equipments are just assemblies of sensors and I/O converters to transform those signals into recognizable data to our human brains. This doesn't mean that the so called matter does actually exist that way.

1.3 The Flow of the Space

1.3.1 The Space Flows in Time Dimension

If there is a despace hole in the space, the space will replenish it sponta-neously. However, the space is 3 dimensional, time is the 4th dimension. The-current becomes into the-past so that the-future can change into the-current. This causes the space to flow in time dimension as a 4-dimensional flow.

Since our world is moving towards the-future continuously, if we take the space of the whole universe as a flowing current, the direction of the space flow does not depend on the *XYZ* axes (the space dimension), but depends on the *T* axis (the time dimension). And the Time axis (*T* axis) is always orthogonal to any other space axes (*XYZ* axes).

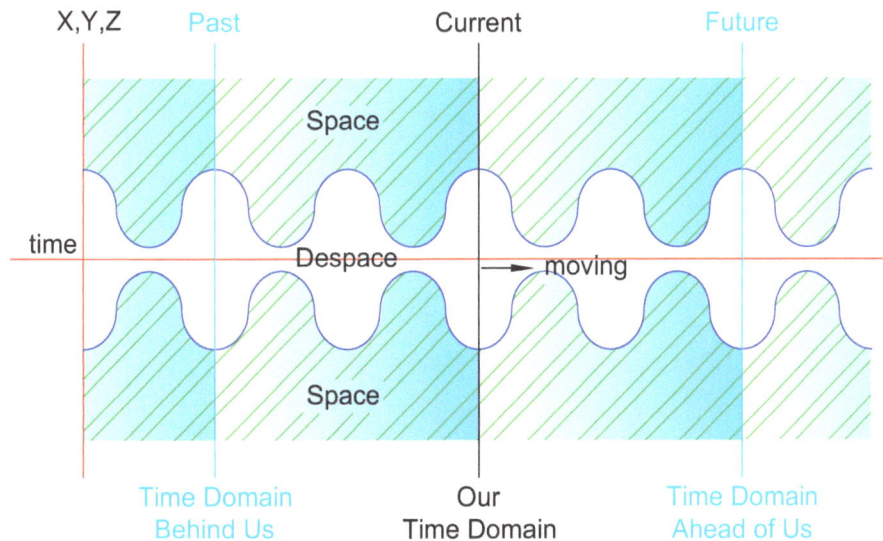

(Fig1.3-1: Despace and time domain)

The space can flow from the-current through the despace hole into the-past. The despace will act as a balloon to distend and shrink repeatedly, to push and pull the space around it as a 4-dimensional waveform with a certain frequency as the *particle-wave* in time-space vibration. This is similar to a vortex in the water.

However, it is not possible for a 3-dimensional device or a creature to detect 4-dimensional things. This is similar to a 2-dimensional plan can not describe a 3-dimensional space.

1.3.2 The Time-Space Vibration

Our universe can be simplified as a section of fluid in a conduit (see Fig1.3-1). However this section of the time conduit includes the *XY*, *YZ* and *ZX* plane. And the axis as the length of the time conduit should be replaced by *T* axis. The space of our world (the-current) is flowing into the-past continuously, at the same time, the-future is also pouring our world with a huge volume of the space continuously.

The space is being dragged into the despace and disappearing into the-past from the-current while the despace is shrinking. In the mean time, the space pouring from the-future into the-current will push away the space around the despace hole while the despace is distending. When these two different opposite forces reach a dynamic balance, an oscillatory balance, the space around the despace hole will become to vibrate periodically.

Additionally, while the space is flowing into the despace hole, the despace hole will act as a vortex. This makes the despace hole to have the spin property.

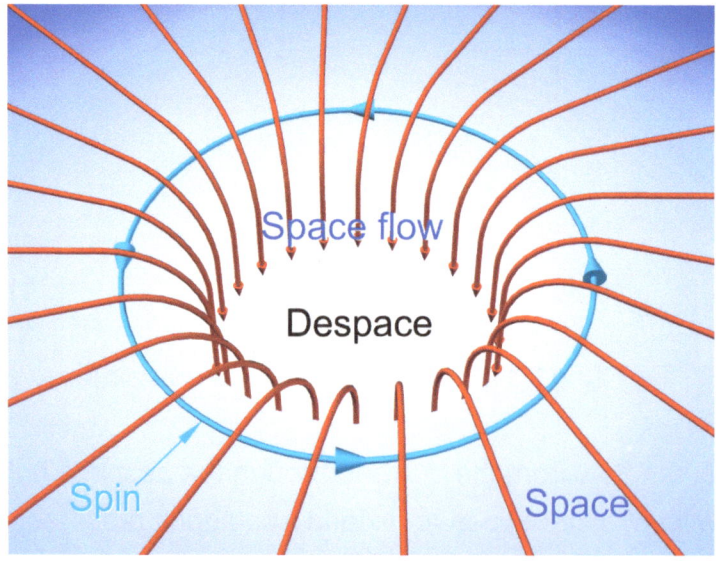

(Fig1.3-2: A despace hole)

While the space around the despace hole vibrates periodically, it will make the space around the despace to act as a balloon to distend and shrink, or to drag and push away the space around. Then the space around the despace can be taken as a 4-dimensional waveform with a certain frequency – *the particle wave*. Note that the space will flow in time dimension after it enters the despace hole. The arrows of the lines inside the despace hole will always point towards the time (*T*) axis.

However, the particle-wave is different from the traditional mechanical wave. The particle-wave propagates in time dimension rather than in space dimensions. So actually, a particle-wave is a 4-dimensional waveform (*XYZ+T*). For any instance, the particle-wave is a 3-dimensional breach of the space. Since the particle-wave generated by the despace is a kind of waveform, it will have a certain period, frequency, amplitude and phase.

1.3.3 The Property of the Space and the Space Flow

Since the space is some kind of the fluid flowing in time-space dimension, we can also define *the Density-Of-Space* (*DOS*) and *the Pressure-Of-Space* (*POS*) in the space.

Though the *DOS* and the *POS* may be nearly constant all over the universe, however, there would be a little difference of the *DOS* and the *POS* where the despace holes exist in the space. And the difference of the *DOS* and the *POS* in the space will cause the space to flow.

1.3.4 The Time Domain

For the despace is a waveform propagates in time dimension, there can't really be a future or a past of our world. The despace wave only exists at an instance of our-current, not the-past nor the-future, because the-past has already gone, and the-future has not yet arrived. The despace has already moved from the-past into the-current, but has not yet moved to the-future.

What we named the-future and the-past in this article is a world ahead of us and a world behind us in *Time-Domain* (see Fig1.3-1). Or we can take them as different sections of worlds in the *Conduit-Of-Time*. And the despace can be taken as a connection between two different worlds. It seems like small black holes connecting between 2 different worlds. The worlds ahead or behind our world could have nothing to do with our world, or maybe similar to our world.

(Fig1.3-3: Despace holes connecting between time domains)

1.4 The Basic Particles

We know that matter is composed of the atoms. The atom is composed of the protons, electrons and neutrons. A neutron is composed of one proton and one electron. To define matter, we have to define the basic particles instead.

1.4.1 Electron

The electron is a particle wave vibrates in time-space dimensions. When the-future (the upstream time domain) is pouring the-current (the current time domain) with space, it will push the space of the-current to move into the-past (the downstream time domain). If the pushing force of the space flow by the-future is not strong enough, the pushed space will rebound from the-past back into the-current or even into the-future. This will cause the space to vibrate in time dimension as a 4-dimensional particle wave, and make the electron to drag and push away the space around its vicinity.

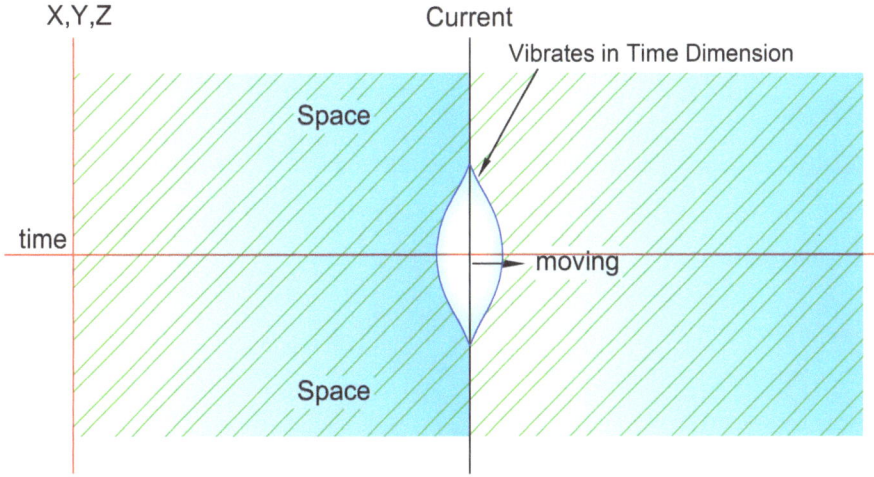

(Fig1.4-1: Time-space vibration of an electron)

However, the electron does not have a despace hole at its center, which means the space will not be dragged and disappear into an electron. The time-space vibration of an electron will be similar to a 4-dimensional sine wave. The space flow cause by an electron will become balanced to zero at a distance away from the electron in every wave cycle.

1.4.2 Neutrino

A neutrino is a very special particle. It could be a despace hole without space vibration and no space flows through it. Or perhaps it could be an end of a despace hole of the world ahead of us in time domain.

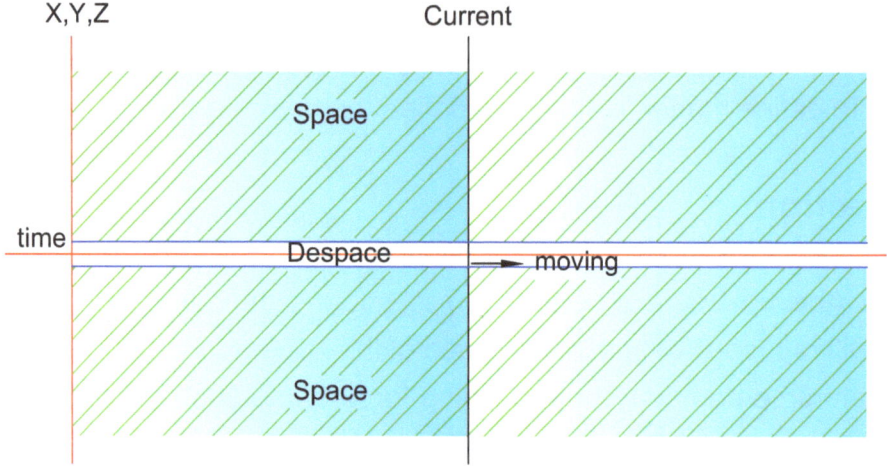

(Fig1.4-2: Time-space vibration of a neutrino)

A neutrino is only a breach of space in the space. There is no time-space vibration of a neutrino. It's like a bubble in the water.

1.4.3 Proton

The proton is a despace hole that the space will flow through it and disappear into the-past. And the edge of despace of a proton will distend and shrink periodically which will push away and pull back its vicinity space periodically.

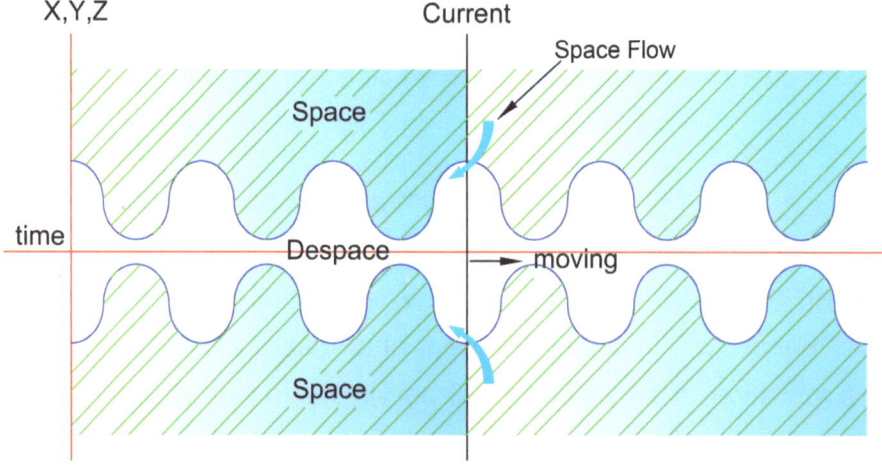

(Fig1.4-3: Time-space vibration of a proton)

The space in the-current is being dragged into the-past. In the mean time, the space is also continuously pouring from the-future. This will cause the space to vibrate in time dimension as a 4-dimensional particle wave. It will drag and push away the space around its vicinity by this vibration.

The time-space vibration of a proton will also be a 4-dimensional sine wave. The space flow cause by the vibration of a proton will become balanced to zero at a distance within every wave cycle. However, the space flow of the space being dragged into the-past from the despace hole of a proton is not able to be balanced. The space will be sucked into the despace of a proton continuously at a constant rate.

1.4.4 The Phase Difference between Proton and Electron
The time-space vibration of a particle can be described as a waveform. It should have the particular amplitude, the frequency and the phase. The amplitude and the frequency of the time-space vibration of a proton and an electron are almost all the same since they do come from the same time-space drag and pushing. However, the phase difference of the time-space vibration between a proton and an electron is just 180°, which means they can counterbalance with each other.

1.4.5 Neutron
A neutron is a combination of a proton and an electron. It can be deemed as a proton and an electron are bounded together revolving around each other in a very short distance. The space vibration outside of a neutron in a certain distance will be almost neutralized.

1.5 Gravitation

Gravitation is equal to everything, for a huge heavy rock or a small light feather, and even for the mass-less photon or the electromagnetic wave, the effect of the gravitational attraction is all the same to everything. The gravitational attraction to an object concerns only with the gravitation field where the object is positioned. This is similar to the objects drifting in the water. Without considering any other forces, for a small leaf and a giant vessel, the acceleration caused by the drag effect of the water flow is all the same to them (If they are in the same outline with respect to the flow of water).

1.5.1 Space Flow Caused by De-spacing

Since Earth is a huge gathering of atoms which contains the protons, the proton contains a despace hole - the small holes of the space in the space. This can be simplified as a giant despace hole in the space. And the space is flowing into the giant despace hole and disappearing.

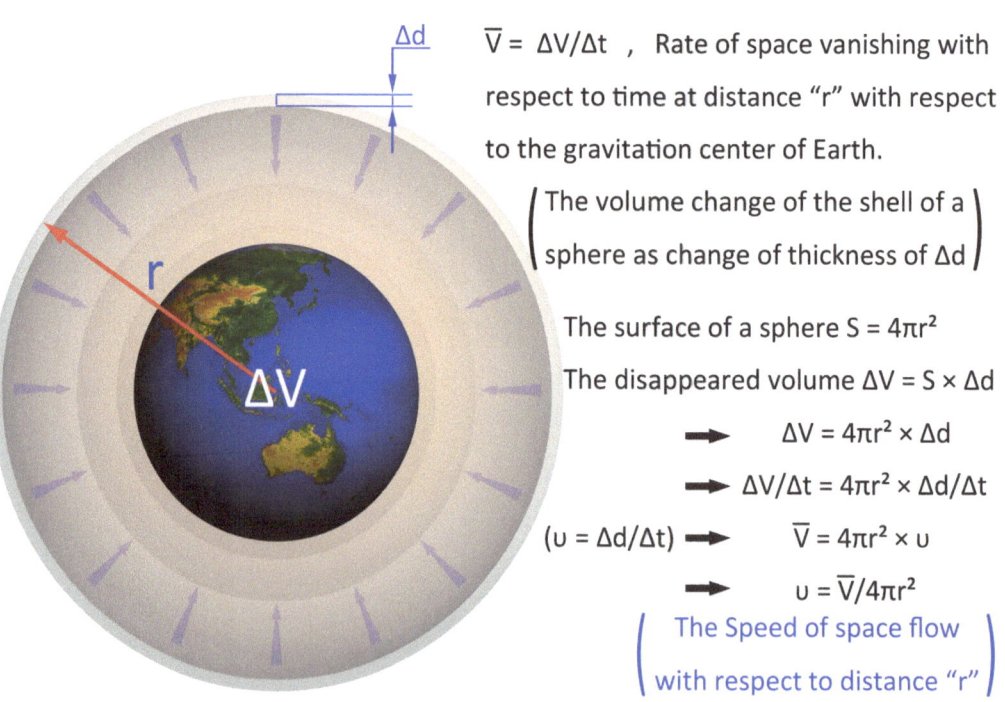

$\overline{V} = \Delta V/\Delta t$, Rate of space vanishing with respect to time at distance "r" with respect to the gravitation center of Earth.

$\left(\begin{array}{l} \text{The volume change of the shell of a} \\ \text{sphere as change of thickness of } \Delta d \end{array} \right)$

The surface of a sphere $S = 4\pi r^2$

The disappeared volume $\Delta V = S \times \Delta d$

$\Delta V = 4\pi r^2 \times \Delta d$

$\Delta V/\Delta t = 4\pi r^2 \times \Delta d/\Delta t$

$(\upsilon = \Delta d/\Delta t)$ → $\overline{V} = 4\pi r^2 \times \upsilon$

$\upsilon = \overline{V}/4\pi r^2$

$\left(\begin{array}{l} \text{The Speed of space flow} \\ \text{with respect to distance "r"} \end{array} \right)$

(Fig1.5-1: Space flow by space disappearing into core of Earth)

Within the flowing of the space, all of the substances which composed of the atoms that drifting in the space will be dragged by the space flow and be accelerated by the space drag even they are stationary with respect to the giant despace hole.

Assume that there is a certain volume of the space ΔV vanishing cause by Earth continuously. As the space is vanishing into the core of the Earth, the space outside of the vanished part will replenish the vanished space spontaneously. This means that the space surrounding the Earth will move towards the center of Earth by the space flow, which will drag all the objects drifting in the space to move towards the center of Earth.

Diminution of the space volume in the core of Earth equals change of the radius of the volume of a shell at a distance r with respect to the center of Earth. The volume of the shell can be described as following when r is very large:

$$\Delta V = S \times \Delta d = 4\pi r^2 \times \Delta d \qquad (\Delta d = \text{the thickness of the shell})$$

$$\rightarrow \overline{V} = \frac{\Delta V}{\Delta t} \qquad \text{(rate of space diminution)}$$

$$\rightarrow \overline{V} = \frac{4\pi r^2 \times \Delta d}{\Delta t} = 4\pi r^2 \times v \qquad , \quad \text{Where } v = \Delta d / \Delta t$$

$$\rightarrow v = \frac{\overline{V}}{4\pi r^2} \qquad (V = \text{the speed of the space flow})$$

The speed v is the speed of the space flow with respect to the center of Earth caused by Gravitation generated by Earth.

1.5.2 Drag Coefficient of Space Drag

Since all of the substances are composed of protons and electrons, the gravitational attraction to all of the substances can be simplified as gravitational attraction to protons and electrons.

The space drag to protons and electrons can be described as a drag equation with a drag coefficient C_d. The acceleration caused by the space drag can be described as below:

$$F = m_p \times g = C_d \times \rho \times v^2 \times A \qquad \text{(i)}$$

(with referring to fluid dynamics)

m_p is the mass of the particle

g is the gravitational acceleration

ρ is the density of the fluid, the space, which equals to the density of the particle with a factor, here we omit this factor

v is the speed of the particle with respect to the space flow

A is the cross-sectional area of the particle perpendicular to the direction of space flow

$$\rightarrow g = \frac{C_d \times \rho \times v^2 \times A}{m_p} \qquad \text{(ii)}$$

$$\to g = \frac{C_d \times \rho \times v \times (v \times \kappa_1) \times \dfrac{1}{\kappa_1} \times A}{m_p} \qquad \text{(iii)}$$

$\Delta d = v \times \kappa_1$, displacement of the particle is proportional to its velocity

κ_1 can also be taken as a factor of the time.

$$\to g = \frac{C_d \times \rho \times v \times (\Delta d) \times \dfrac{1}{\kappa_1} \times A}{m_p} \qquad \text{(iv)}$$

$$\to g = \frac{C_d \times \rho \times v \times (\Delta d \times A \times \kappa_2) \times \dfrac{1}{\kappa_2} \times \dfrac{1}{\kappa_1}}{m_p} \qquad \text{(v)}$$

$V = \Delta d \times A \times \kappa_2$, the volume of the particle equals to the cylindrical volume with a factor.

$$\to g = \frac{C_d \times \rho \times v \times (V) \times \dfrac{1}{\kappa_2} \times \dfrac{1}{\kappa_1}}{m_p} = \frac{C_d}{\kappa_1 \kappa_2} \cdot \frac{\rho \times v \times V}{m_p} \qquad \text{(vi)}$$

$C_x = \dfrac{C_d}{\kappa_1 \kappa_2}$ is the simplified drag coefficient.

$$\rightarrow g = C_x \cdot \frac{\rho \times v \times V}{m_p} \qquad\qquad\text{(vii)}$$

$m_p = \rho \times V$, mass = density × volume

$$\rightarrow g = C_x \cdot v \cdot \frac{\rho \times V}{m_p} = C_x \cdot v \cdot \frac{m_p}{m_p} \qquad\qquad\text{(viii)}$$

$$\rightarrow g = C_x \cdot v \qquad\qquad\text{(ix)}$$

As a conclusion, the acceleration to the particles caused by the space drag is proportional to the speed of the space flow.

Since all of the substances are composed of the protons which are the despace holes in the space. And the despace holes would de-space (deleting the space) all the time and cause the space to flow. The flow of the space will cause acceleration to the particles when the particles drifting in the space are dragged by the space flow. This is the so called *Gravitational Acceleration*. And the planet which composed of protons will generate Gravitation consequently. Thus Gravitation concerns only with the quantity of protons of the planet.

1.5.3 Weight Effect

For the water in the water, no matter how the water flows, the water is still water, nothing will change. But if there are bubbles in the water, the flow of water will cause the bubbles to move along with the water. The bubbles will hence have acceleration and kinetic energy. If the bubble will not affect other bubbles in any way except for collision, the bubbles can pass through bubbles freely.

For the despace holes in the space, they are very similar to the bubbles in the water. However, different despace holes have different effect to the space flow and vibration.

For a neutrino, it's just the same case with a bubble in the water. It can pass through the substances freely and does not affect anything. It may also be dragged into the despace hole and disappear.

For an electron, if it's very closed to other particles, its time-space vibration will affect the particles in its vicinity and cause the other particles to move and gain some kinetic energy. This is the so called the *Weight Effect*. It's just the action and reaction, and hence makes the electron to have weight property.

Only the particles with despace holes which can de-space can truly generate Gravitation, and have genuine weight property.

1.5.4 Gravitational Wave

Since Earth or other celestial objects are compose of protons. And all the protons have the same frequency and phase. It can be taken as Earth is one big proton which can de-space and makes the space to flow towards the center of this big proton.

The strength of the space flow (or the amplitude of the despace wave) complies with the cycle of the despace wave. This will make the flow of the space to become as a kind of waveform, a longitudinal waveform. This is the so called *Gravitational Wave*. And the speed of Gravitational Wave is just the speed of the space flow, and it's inversely proportional to the square of the distance.

1.6 Electric Field

The most interesting phenomenon of electrostatic force is that the force between two electric charges with different properties is attraction, and the force between two electric charges with the same property is repulsion, and the force of an electric charge to anything with electrical neutrality is zero. How come such a force could be so funny and selective? What is our definition of a zero force?

1.6.1 Time-space Vibration vs. Electric Property

The assumption of the phase difference of the time-space vibration between a proton and an electron is based on the principle of electric property. I think only the same things with reverse properties to each other can neutralize or counterbalance with each other.

Assume the phase difference of the time-space vibration between a proton and an electron is about $180°$. As they are two similar waveforms with phase difference of $180°$, they will have a destructive interference when they are placed together, which means the time-space vibration of them can be counterbalanced.

1.6.2 Distention and Shrinkage of Space

Basically, the distention of space will push away anything around, and the shrinkage of space will drag everything closer. In other words, the protons and the electrons will always push away and pull back their vicinity space repeatedly. This will push away and draw back the nearby particles with respect to the proton or the electron repeatedly.

Although the forces of distention and shrinkage of space can be counterbalanced with each other, but the effects of distention and shrinkage of the space to the space are not the same even their strength (the amplitude of the time-space vibration) is the same.

As there is the *Pressure-Of-Space (POS)* within the space, when a particle is distending the space, the positive pressure or the strength of pushing away the space can be as great as possible. However, for the shrinkage of space of a particle, the maximum negative pressure of pulling the space cannot be greater than -*POS* with respect to the local *POS* of the space.

This is the same of a compressor with respect to a vacuum generator in the general atmosphere. The pressure of the compressed air can be as high as possible regardless of the pressure of the atmosphere. However, the maximum negative pressure can only be the vacuum.

1.6.3 Attraction between Different Electric Properties

When two particles with above characteristics of distension and shrinkage of space but opposite to each other are placed together, which means both particles have 180° phase difference between them. We can denote the two particles as **A** and **B**.

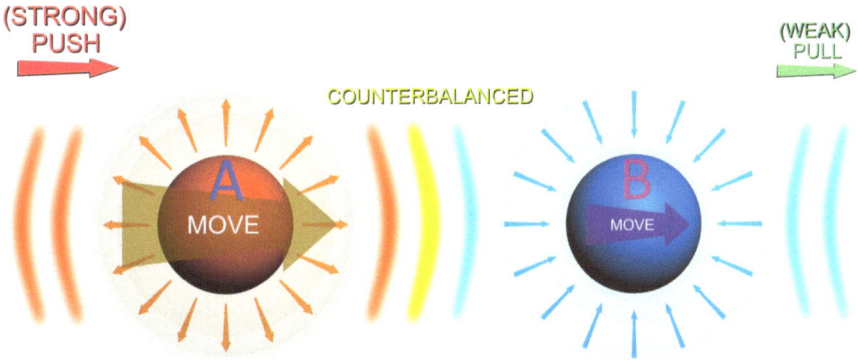

(Fig1.6-1: **A** is distending and **B** is shrinking)

Assume their phase is between 0° to 180°. In this case, when **A** is pushing away its vicinity space, and **B** is pulling back its vicinity space at the same time, if **A** and **B** are close enough, the space vibration between them will be counterbalanced. But the space at the outer side of **A** will push **A** towards **B**, and the space at the outer side of **B** may pull **B** towards its outer side a little. However, these two forces on **A** and **B** are not equal. The force of pushing is stronger than pulling. This will make **A** and **B** to get approaching. Meanwhile, this will also make the mass center of **A** and **B** to move rightwards a little bit.

After a half cycle of the particle vibration, which means their phase is between 180° to 360°, the space vibration of **A** and **B** will be reversed. This means **A** is pulling back its vicinity space, and **B** is pushing away its vicinity space at the same time. Then the space vibration between A and B will still be counterbalanced, but the space at the outer side of B will push B towards A, and the space at the outer side of **A** may pull **A** towards its outer side a little. This will make **A** and **B** to get approaching, and also make the mass center of **A** and **B** to move leftwards a little bit.

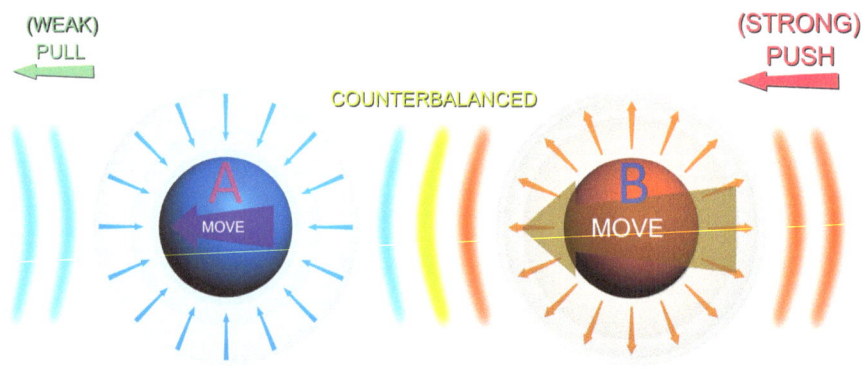

(Fig1.6-2: **A** is shrinking and **B** is distending)

The combination movement of **A** and **B** will then become an attraction force between them. Moreover, since there is always one of them pushing away the other one, they can never get real contact with each other. The two particles will keep vibrating back and forth with respect to their mass center, and the distance between them will also increase and decrease repeatedly. Since the all the particles have a spin property, the attraction between **A** and **B** will then make them to revolve around their mass center.

Normally, **A** and **B** would be a proton and an electron, more exactly, a nucleus and several electrons. Since the mass of a nucleus is much greater than an electron, and the nucleus will spin continuously, this will cause the electrons to revolve around the nucleus in elliptical courses.

1.6.4 Repulsion of the Same Electric Property

When two particles with the same charges are placed together, which means both particles have no phase difference between them. Assume their phase is between 0° to 180°. In this case, when both particles are pushing away their vicinity space, the space increased between them will become twice of the space at their outer side, thus the two particles will push away each other.

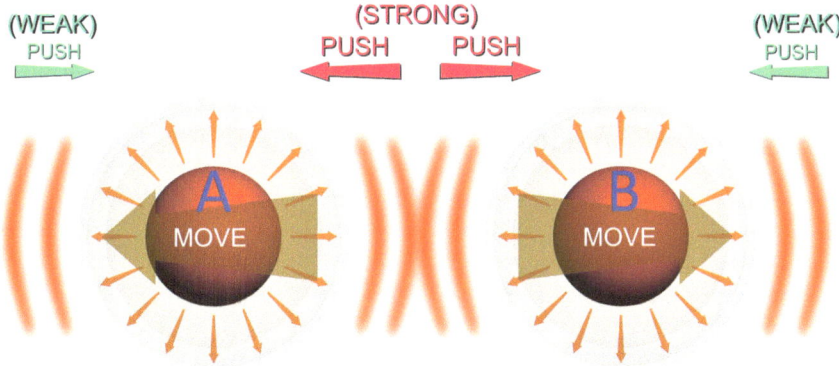

(Fig1.6-3: **A** and **B** are both distending)

When the phase of both particles is between 180° to 360°, the space vibration of them will be reversed, which means both particles are pulling back their vicinity space. This will cause both of them to get closer. However, the force of pulling at this stage is smaller than the force of pushing at last stage. The combination movement of both particles will then become a repulsion force between them.

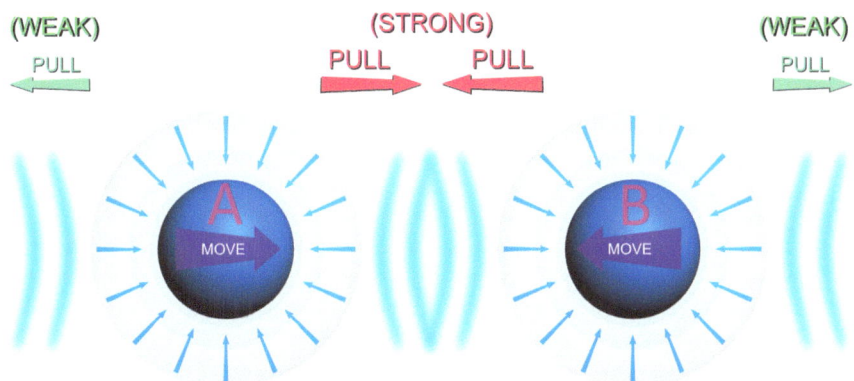

(Fig1.6-4: **A** and **B** are both shrinking)

1.6.5 Electrostatic Force to an Electrical Neutrality Object

When the number of protons and electrons are balanced in a substance or an atom, the space vibration of the particles which it contains will then be neutralized by itself with a certain distance away from it, and attain to a state of *Electrical Neutrality*. If the number of protons and electrons are not balanced in a substance or an atom, the space vibration of the particles which it contains can't be totally neutralized. The rest of the unbalanced electrical charges will then determine the charge property of the substance.

When an electrical neutrality object is placed next to an electrical charge, because the average space vibration of the electrical charge within every cycle is zero, and the frequency of the vibration is so high, though even the particles of the object might be oscillated by this space vibration, but the whole object will always seem to be static.

However, if the electric field of the electrical charge is so strong and the object is placed close enough to the electric field, the electrons in the object will then be affected to be attracted or repulsed. This will cause the electrical neutrality object to lose electrical neutrality.

1.6.6 Electric Field with respect to Distance

The un-counterbalanced space vibration of an electrical charge will radiate as a sphere *Longitudinal Wave*, because the direction of its space vibration is parallel to the wave propagation.

Since the space vibration of a particle is a sphere, the acceleration cause by the space vibration will also be inversely proportional to the square of the distance from the center of the electric field.

1.7 Magnetic Field

The phenomenon of the magnetic field is quite similar to the electric field. The force between two of the magnetism with different polarity is attraction, and the force between two of the magnetism with the same polarity is repulsion, and the magnetic force to any non-magnetic material is zero.

The magnetic force could be so strong to make a train to float in the air. Since it's so strong, how come we never feel about it? We denote the north pole of Geomagnetic as **N** and the south pole of Geomagnetic as **S**. Is there really a fixed direction of magnetism?

1.7.1 Orbital Movement vs. Space Vibration of the Electron

We know that the magnetism must have something to do with the orbital movement of an electron around the nucleus. When an electron is moving on its orbit, it can be taken as the electron is moving back and forth when we look at it by side. At the same time, the electron is also vibrating its vicinity space spherically.

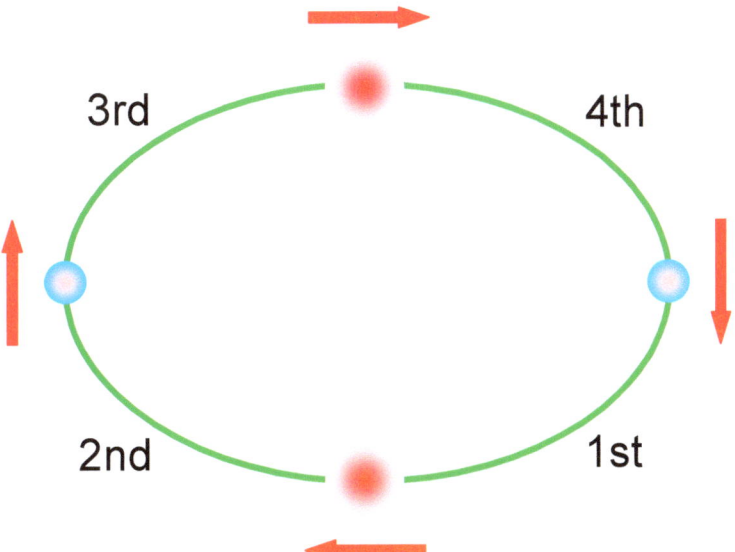

(Fig1.7-1: Orbital movement with respect to time-space vibration of an electron)

The orbital movement of the electron has a fixed period τ, and the space vibration of the electron also has a fixed period ϕ. If both periods relate to the electron have a special relationship ($\tau = 2^n \times \phi$, $n \in N$), and the revolution speed of the electrons reaches a certain level, the electron will generate the magnetic field.

We will consider the condition of $\tau = 2\phi$, two of space vibration cycle in one revolution cycle of an electron around the nucleus in the following. Which means there will be 180° phase change of the space vibration in every quarter of the revolution course of the electron around its nucleus. And separate the revolution course of the electron around its nucleus into 4 quarters as below:

When an electron is moving leftward in the 1st quarter of the revolution course and it's also distending its vicinity space with the phase between 0~180°, then the space increased at the left side of the electron will be twice or more than which at the right side. The space vibration caused by the electron can be taken as the space is being pushing leftward.

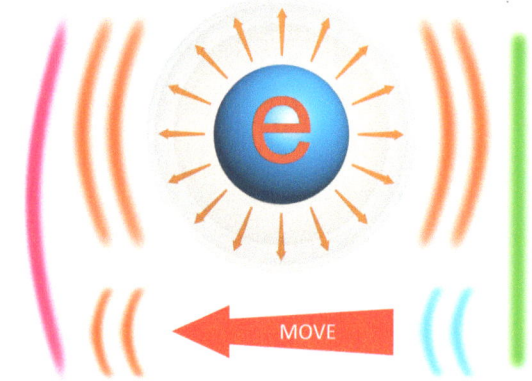

(Fig1.7-2: Orbital movement 1st quarter, electron distending)

After the electron passes into the 2nd quarter of the revolution course, the phase angle of the electron particle wave also passes over 180°. That means the electron is starting to shrink its vicinity space. Then the space decreased at the right side of the electron will be twice or more than which at the left side. The space vibration caused by the electron can be taken as the space is being pulling leftward.

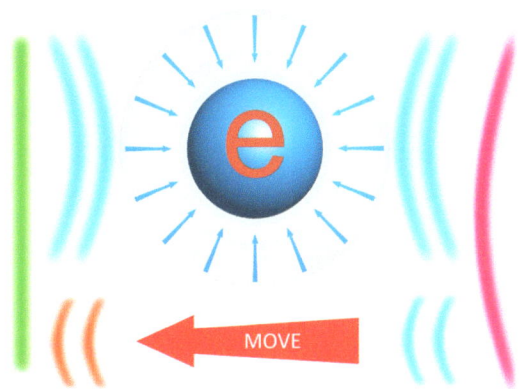

(Fig1.7-3: Orbital movement 2nd quarter, electron shrinking)

Similarly, for 3rd and 4th quarter of the revolution course of the electron, the situation is all the same except the direction should be changed to rightward. Combine the 4 stages of the space vibration as below. As a result, the space will vibrate back and forth alternately by the orbital movement of the electron. And the average of space movement of every cycle will be zero.

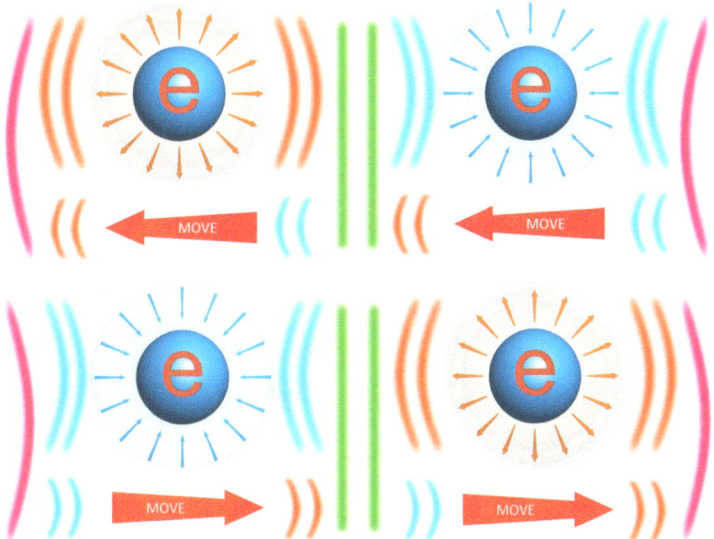

(Fig1.7-4: Combined space vibration with τ = 2φ)

However, this is a very rough idea of the space vibration vs. the electron movement. Actually, the revolution speed of an electron around the nucleus is not constant, and its orbit is much similar to an ellipse. This is why the result will become a back-and-forth space vibration.

1.7.2 Magnetic Field Generated by Space Vibration

When a material contains the atoms with the electrons moving in their orbits, most of the space vibration cause by the electrons movement could be counterbalanced. If the space vibration after self counterbalanced of the electrons in an atom becomes alternately vibrating in one direction, and the other atoms also vibrate the space in the same direction synchronously. We can take as the whole material is vibrating the space alternately in one direction.

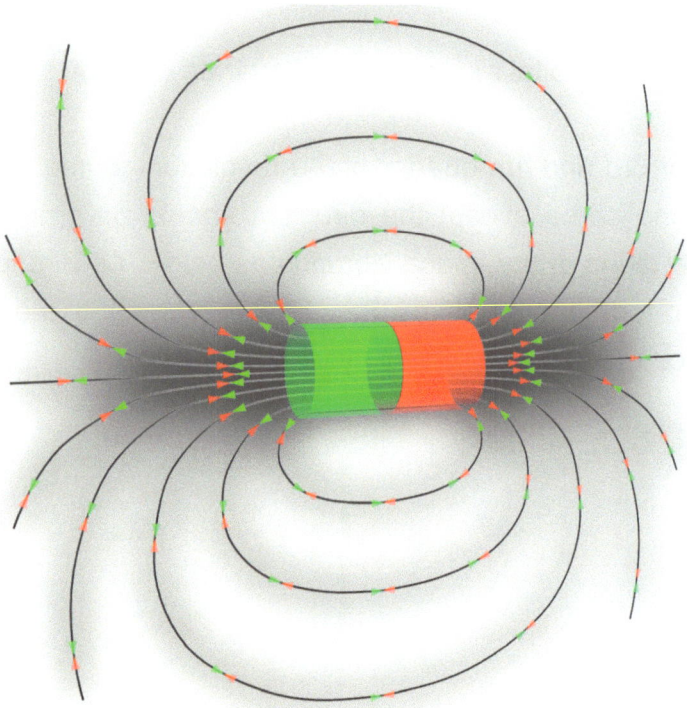

(Fig1.7-5: Magnetic field caused by space vibration)

Thus the material will generate the magnetic field which is an alternate vibration of the space. And the magnetic line of force will be bended when it leads outside of the material. That's because the space increased at one direction will try to fill the space decreased at the reverse direction.

Since it's vibrating so fast, you would not be able to sense about it. And the average of the space vibration will always be zero. A non-magnetic material will seem not be affected by it.

1.7.3 Reaction between Magnets

If we place two of the magnets close together in series, if the magnetic poles of them are different in the middle of them, the space vibration between them will be neutralized, and the outer side of each other magnet will push one towards another alternately. This will become an average attraction force of the magnets. (Please refer to *Part 1.6.3* for the same reason).

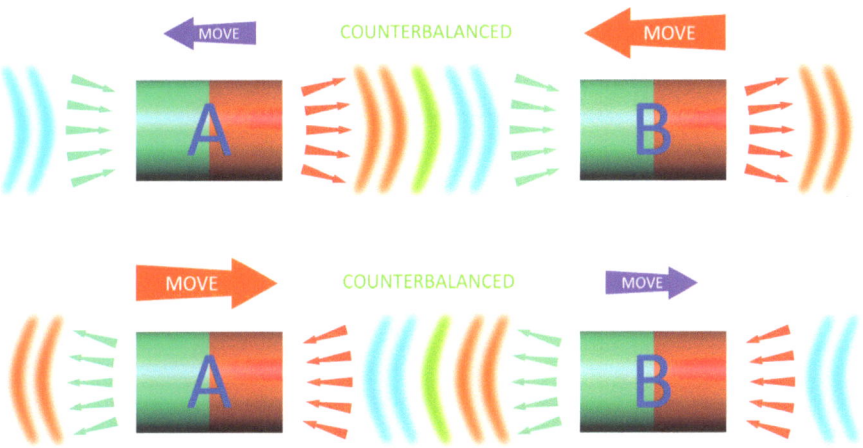

(Fig1.7-6: Attraction of magnetic field)

If the magnets are placed in the opposite way, that means the magnetic poles in the middle of the two magnets are the same, then the space vibration between them will pushing them away from each other when they are increasing space towards the middle of them.

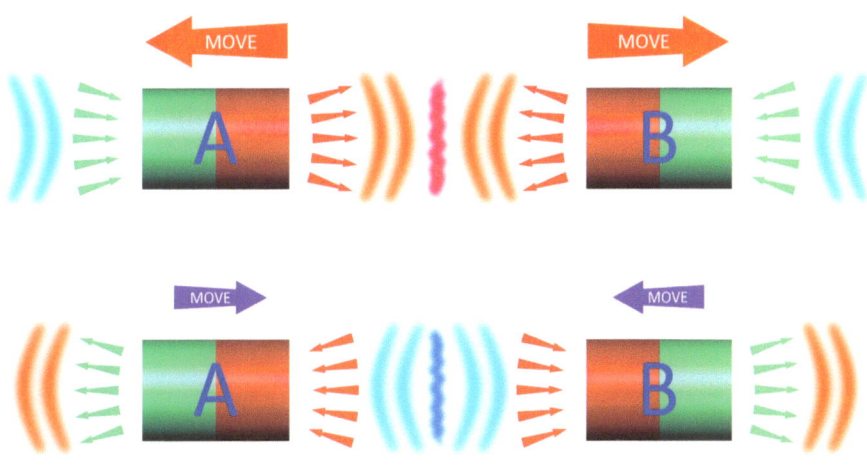

(Fig1.7-7: Repulsion of magnetic field)

Though the magnets will try to pull back each other after pushing away, but the force of pulling back is much smaller. This will become an average repulsion force of the magnets. (Please refer to *Part 1.6.4* for the same reason).

1.7.4 Magnetic Poles

The magnetic pole of Geomagnetic is only a reference for the other magnetisms. The magnetisms may not have to be truly synchronized to the Geomagnetic. We can only compare them by the force of attraction or repulsion. So there is not actually a mono-direction of any pole but an alternate one.

Meanwhile, if the magnetic force is a force one way out, it should affect everything. If the direction of the magnetic force is a mono-direction, there should be three different results when we place 2 magnets in series as below:

A) Repulsion   :   **M→ ←M**

B) Attraction   :   **M← →M**

C) Neutralized   :   **M← ←M** or **M→ →M**

Obviously, this contravenes the fact.

1.7.5 EMF (electromotive force)

Since the direction of the space vibration of the magnetic wave is parallel to the wave propagation. The magnetic wave is also a *Longitudinal Wave*.

Whenever we place a conductor in a magnetic field, the electrons in the conductor will move synchronously with the space vibration caused by the magnetism. If the magnetism is changing its direction, the direction of the electron movement will change correspondingly. This will produce an *Electromotive Force*, and forces the electrons to move towards a particular direction, which creates the electric current.

1.8 The Basic Construction of an Atom Nucleus

The protons are always repulsive to other protons. How can they be bound together in an atomic nucleus? Meanwhile, atomic nucleuses are also repulsive to other atomic nucleuses. How can they form a molecule?

In modern physics, physicians found some of subatomic particles as *Mesons* and *Gluons* which are composed of one *quark* and one *anti-quark*, and the *Strong Force* of quark interaction holds quarks and gluons together to form protons, neutrons and other particles.

However, except for the neutron, the mean lifetime of a subatomic particle outside of the nucleus is less than 10^{-5} second, and the shortest one could be less than 10^{-28} second. If a nucleus is formed by those subatomic particles, which means those subatomic particles are inside of the nucleus, thus why they can't exist outside of the nucleus?

Moreover, more and more different subatomic particles are found in recent year due to improvement of equipments and sophistication. The so called *Standard Model* has been modified for times. So will there be an end?

1.8.1 A Basic Model of a Hydrogen Atom

According to what I've described in *Part 1.4* about the space vibration of the particles and *Part 1.6* about the electrostatic force, we can build a simplest model of a deuterium atom nucleus for example.

Deuterium, also called heavy hydrogen, is a stable isotope of hydrogen. Deuterium is frequently represented by the chemical symbol D or ^2H. A deuterium atom contains of an electron, a proton and a neutron. Since a neutron is composed of one electron and one proton, thus we can describe as two electrons and two protons in a deuterium atom.

However in general conditions, a deuterium atom will always catch another hydrogen atom to form a heavy water molecule. They will share their orbital electrons with each other for a balance. This means they will throw out their orbital electrons and accept the one from the other atom. In this case, the deuterium atom will not loss an electron, but act as gaining one extra electron instead.

Thus, a deuterium atom can be taken as composed of three electrons and two protons. And we can make the model as following. (Note that the outside electrons are shared electrons).

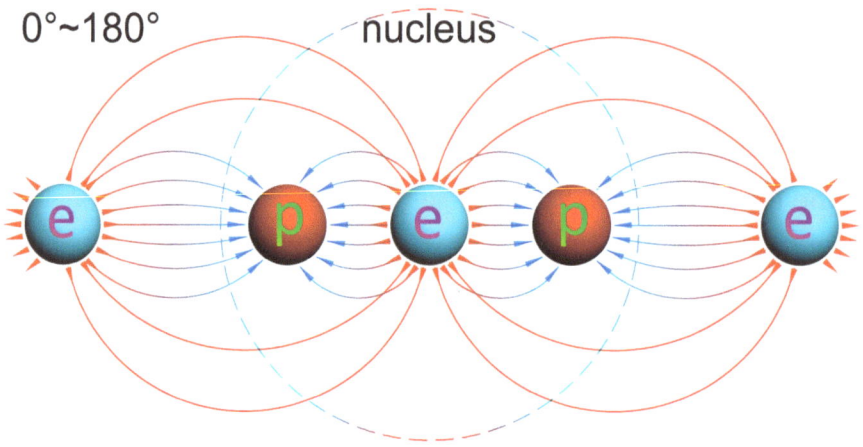

(Fig1.8-1: A deuterium atom, phase 0~180°)

Assume that the phase of all the particles are synchronous and between 0~180°. During this stage, the electrons are distending and the protons are shrinking the space. The protons will pull the orbital electrons closer, but the orbital electrons will push each other away. The orbital electrons will be kept in their orbit when the forces are balanced, and the protons will be pushed towards the center of the nucleus.

Though the electron inside the nucleus will push the protons away, but since the space is flowing into the proton, there will not be enough "space" between them for an electron to distend the space, and the protons are pulling both closer at the same time, thus the nucleus can be so small and crowded.

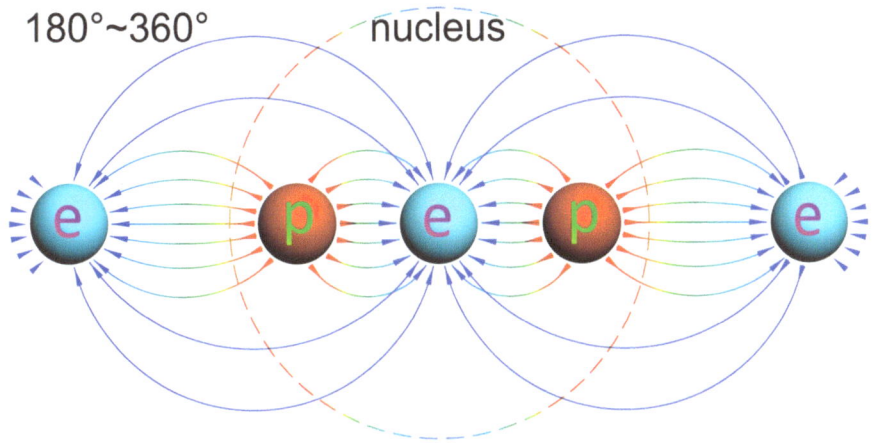

(Fig1.8-2: A deuterium atom, phase 180~360°)

Once the phase of the particle vibrations passes over 180°, all the space vibrations will be reversed. During this stage, the electrons are shrinking and the protons are distending the space.

The protons will push the orbital electrons away, but the orbital electrons will try to get closer. The orbital electrons will be kept in their orbit when the forces are balanced.

Though the protons are pushing everyone away, but the electron in between them will pull them closer. In the mean time, since the space is flowing into the protons, the protons and the inner electron are too close that the protons cannot have enough "space" to distend to push each other away in middle of them. Thus the nucleus can still be kept as small and crowded.

1.8.2 The Mass of a Proton in the Nucleus

When the protons and the electrons are concentrated within the nucleus, the distension of the electron of the space will increase the speed of space flow into the proton, but the shrinkage of the electron of the space will only decrease a little speed of space flow. This will make the protons to become heavier than a single one outside of the nucleus. The larger the nucleus is, the heavier the protons will be.

1.8.3 Orbital Electrons and Formation of Molecules

At any moment, the orbital electrons will be pushed away and pulled close with respect to the nucleus center, and be kept within a range. The atoms which share with the orbital electrons will be bound by them as well and form a molecule. And the molecules which share with their orbital electrons will also be bound together to form a polymer or a material.

The orbital electrons can never get close into the nucleus and also not able to escape from the nucleus. They will be bound in between the nucleuses and revolve around the nucleuses. And the electron inside of the nucleus will also revolve around a center point inside the nucleus.

1.8.4 Radioactive Decay

When a neutron was emitted out of the nucleus, there will be no more pressing force to put the proton and the electron together. They are going to revolve around their mass center. However, the proton is a despace hole, and the electron is almost massless with respect to the proton. This will make the electron to revolve around the proton.

When the neutron escapes out of the nucleus, the radius of the revolution orbit of the electron around the proton will become getting larger. Meanwhile, the electrostatic force will try to maintain the angular speed of the electron revolves around the proton. This will increase the revolution speed of the electron.

However, the revolution speed of the electron is very close to light speed originally. This will make the revolution speed of the electron to become higher than light speed and induce the **Cherenkov Radiation**. (*Please refer to chapter 2, Part 2.2.6*), and generates very high frequency photons such as gamma rays. Finally, the neutron will decay into a proton and an electron and emit gamma rays. As a result, the whole substance will become lighter (or lose mass) and generate nuclear energy.

1.9 Anti-Matter

The matter is composed of the particles such as protons and electrons. The *Anti-Matter* is composed of the *anti-particles* such as *anti-protons* and *positrons*. To define Anti-matter, we should define anti-particles instead.

1.9.1 Electric Charge

An anti-proton is totally the same with a proton except a proton carries positive charge but an anti-proton carries negative charge. Similarly, a positron is totally the same with an electron except an electron carries negative charge but a positron carries positive charge.

However, the definition of the electrical charge is based on the charge property of a proton as a reference. We define the charge which a proton carries as +1e. Meanwhile, an electron carries a reverse charge with respect to a proton, thus the charge which an electron carries should be -1e. So actually, there is not an absolute value of electric charge. It's only a reference value based on a proton.

1.9.2 Anti-proton

An anti-proton is a proton which carries -1e electric charge. In this case, the electrostatic force is attractive between an anti-proton and a proton, and it's repulsive between an anti-proton and an electron, and it's also attractive between an anti-proton and a positron. This can be explained by the demonstrations in *Part 1.6.*

For the despace vibration, the space vibration of an anti-proton is all the same with a proton, the center of an anti-proton is also a despace hole. Thus an anti-proton will also cause space flow and generate gravitational acceleration. However, the phase difference of the space vibration between an anti-proton and a proton is 180°. When the proton is distending the space, the anti-proton is shrinking; when the proton is shrinking the space, the anti-proton is distending.

1.9.3 Positron

A positron is an electron which carries +1e electrical charge. In this case, the electrostatic force is attractive between a positron and an electron, and it's repulsive between a positron and a proton.

The space vibration of a positron is all the same with an electron, there is no despace hole in the center of a positron. The space vibration of a positron will also cause the Weight Effect. However, the phase difference of the space vibration between a positron and an electron is 180°. When the electron is distending the space, the positron is shrinking; when the electron is shrinking the space, the positron is distending.

As a conclusion, the difference between a particle and an anti-particle is that their phase difference of the time-space vibration is 180°.

1.9.4 Anti-matter

For anti-matter composed of anti-particles compare with matter, they are actually all the same in every way. This means the characteristics or the specification of anti-matter to anti-matter is all the same with matter to matter.

On a planet composed of anti-matter, then what the anti-matter does is just all the same with the matter on a planet composed of matter. In this case, the matter on the planet as Earth composed of matter can be taken as anti-matter with respect to the matter on the planet composed of anti-matter. On the point of space vibration, the planet composed of anti-matter is a planet whose composition particles have 180° phase difference of the time-space vibration to those on Earth.

In other words, what we defined as anti-matter is not something opposite to matter, but the same thing which carries reverse charge. ***Anti-matter is not possible to cause Anti-Gravitation***. What truly opposite to matter is the space.

1.10 Annihilation

Annihilation is defined as "total destruction" or "complete obliteration" of an object. In physics, the word is used to denote the process that occurs when a subatomic particle collides with its respective antiparticle.

Annihilation occurs when an electron and a positron collide, and when a proton and an anti-proton collide, but not for an electron with a proton, nor a positron with an anti-proton.

Annihilation of particles and anti-particles will cause matter to loss mass and transform mass into energy and emit light or electromagnetic waves. General annihilation can be demonstrated with annihilation of a proton and an anti-proton, annihilation of an electron and a positron.

1.10.1 Annihilation of a Proton and an Anti-proton

Since the electric charge which a proton carries is +1e, and -1e for an anti-proton, the electrostatic force between a proton and an anti-proton will be attractive. Thus when a proton and an anti-proton are placed close enough, they will be bound together by the attraction force.

The annihilation of a proton and an anti-proton can be simplified as reaction between two of despace holes with 180° phase difference. Since the space is flowing into the despace holes and disappearing, between a proton and an anti-proton, there will be insufficient space for them to distend. Thus, the proton and the anti-proton can get close and overlap with each other almost without any limit.

Since the proton and the anti-proton are despace holes of the space, if they overlap together, they will converge into a new despace hole. However, the despace vibration of a proton and an anti-proton is always reverse to each other. The space vibration will be neutralized where they overlap together and not neutralized where they do not overlap.

Thus the proton and the anti-proton will create interference of space vibration in between them. And this space vibration will become as a new particle and an anti-particle at the same time. The created particle and anti-particle will become attractive to each other and change into a photon and emits. (*Please refer to Chapter 2*).

The number of the photons created depends on the number of interference created. This also depends on the size of the despace hole. The annihilation of a proton and an anti-proton can be taken as two despace holes converge into a new one and lose energy. The quantity of lost energy equals to the number of photons created.

1.10.2 Annihilation of an Electron and a Positron

Since the electric charge which an electron carries is -1e, and +1e for a positron, the electrostatic force between an electron and a positron will be attractive. Thus when an electron and a positron are placed close enough, they will be bound together by the attraction force.

However, the electron and the positron can never get real contact with each other. This is the same between a proton and an electron. Because the despace vibration of an electron and a positron is always reverse to each other. There will always be one of them distending the space in between them. They will be bound together but never get contact to each other.

Since the electron and the positron is spinning all the time, when they are bound together, the combination of attraction force and self spinning will cause the electron and the positron to revolve around their mass center. When their revolution speed around their mass center reaches a certain level, the pair of the electron and the positron will change into a photon. (*Please refer to Chapter 2*).

If the electron and the positron are accelerated fast enough before the annihilation, they will be more accelerated during getting close to each other and let their speed to become faster than light speed. This will cause the electron and the positron to create additional photons due to **Cherenkov Radiation**. (*Please refer to chapter 2, Part 2.2.6*).

1.10.3 Matter is No Matter

As a conclusion, the annihilation of a proton and an anti-proton is different from which of an electron and a positron. The annihilation of a proton and an anti-proton will transmute mass into energy, transmute particles into photons. However, the annihilation of an electron and a positron does not really change anything, but two of the time-space vibration to be bound together and emits as a photon.

If matter is so concrete and tangible, how can two particles annihilate and transmute into formless energy? Or matter is a kind of energy also? Thus, matter is no matter.

If attraction of electrostatic force is the reason for annihilation of particles, why annihilation does not happen to a proton and an electron? Why the electron can never get into the nucleus or hit the proton? Or it is not a particle attracting to another particle. Thus attraction is not attraction.

1.11 Subatomic Particles

Subatomic particles are all particles which are "smaller" than atoms such as quarks, leptons, mesons, hadrons, baryons, gluons, neutrinos and etc. They are all artificial. They do not exist in the Nature. Most of the particles that have been discovered are encountered in cosmic rays interacting with matter and are produced by scattering processes in particle accelerators or Large Hadron Collider (LHC). There are hundreds of known subatomic particles.

The lifetime of these particles is on the order of 10^{-23} seconds. Most of the subatomic particles do not exist individually and is unstable in the Nature. They will decay in a way of emitting radiation in the form of particles or electromagnetic waves and finally disappear.

1.11.1 Subatomic Particles Created by High Energy Collision

When two of protons or atoms are over accelerated and collide with each other, in this case, their relative velocity and momentum with respect to each other will be doubled and become great enough to overcome the repulsion force between them, and hence they are possible to get contact with each other.

The protons are despace holes in the space. When they make contact with each other, the two despace holes will converge into a new one bigger single hole or several smaller holes and cause the radioactive decay which is the same process of the annihilation of a proton and an anti-proton (*see Part 1.10.1*). The frequency and phase of the time-space vibration of the new despace holes may different from the proton, and hence cause the new particles to have different charge property (*Charge Color*) with respect to a proton.

1.11.2 Decay of Subatomic Particles

In an area where the *Density-Of-Space* (*DOS*) is constant, the average size of the despace hole of a proton should also be constant.

Radioactive decay of a subatomic particle occurs when a despace hole in the space is too large or too small, and the *DOS* of the surrounding space will also affect the process of radioactive decay or induce it to happen.

If the despace hole is larger than what it should be in the space, the space will try to fill it to its normal size. Thus the despace hole will become heavier than a proton since the speed of the space flowing into the despace hole would be accelerated and higher than the speed of the space flow caused by a single proton. If the despace hole is too small, it will be lighter than a proton, and it will be filled up immediately, and the despace hole would vanish in the space eventually.

1.11.3 A Particle Is Not a Particle
If the particle is so tangible, how can it change into another one or disappear and release energy? Or the particle is a just wave packet of energy? Thus a particle is not a particle.

1.12 Black Hole and White Hole

A black hole is a region of space in which the gravitational field is so powerful that nothing, not even electromagnetic radiation (e.g. visible light), can escape its pull after having fallen past its event horizon. It is called "black" because it absorbs all the light that hits it, reflecting nothing, just like a perfect blackbody in thermodynamics.

A white hole is the theoretical time reversal of a black hole. While a black hole acts as a vacuum, drawing in any matter that crosses the event horizon, a white hole acts as a source that ejects matter from its event horizon.

1.12.1 Black Hole & White Hole Caused by Despace Convergence

On the huge mass planet where the Gravitation is so powerful which will cause the nucleus of an atom to become very large and contain more protons than what it can do normally. This will cause the atom to become very heavy and the space flow into the despace of the protons will be over accelerated. When the speed of space flow into the despace of the protons is faster than the speed of space flow from the-future, the protons inside the nucleus will become to getting closer and will finally start to converge.

When the protons inside of a nucleus converge into a bigger one, the speed of space flow will become faster at the instant of the convergence. The speed of space flow is supposed to slow down after the convergence of the protons. However, the other nucleuses around the converged one are also converged, the speed of space flow of the whole region will be highly accelerated and cause the converged despace holes to converge again. Finally, all the despace holes will converge into a single one, a giant despace hole in the space, the back hole.

A black hole will suck the space into it and exhaust the space into another world behind our world. The black hole could pass through several sections of worlds and reach the last one as a white hole.

The connection between a black hole and a white hole is like a 4-dimensional tunnel, the black hole is the entrance and the white hole is the exit.

On the surface of the tunnel of a black hole, the direction of the space flow is always parallel to the surface of the tunnel. Thus the despace hole of matter could not be maintained, because the space will fill up the despace hole while no space will flow through the despace, and the despace will no longer exist. In other words, the substance could not exist in a black hole.

After all, at the exit of a white hole, there is no substance being throwing out, but the space being pouring to that world. This will generate the *Anti-Gravitational* force from the center of a white hole. Because the direction of space flow of a white hole is reverse to which from a star.

1.12.2 Anti-Gravitational Particle

Since the Future is pouring our world with space, the space must have flowed through a special particle from the Future to our world. Meanwhile, since the space is flowing into the despace holes from our world into the Past, we can conclude that the particle which pouring our world with space must be a despace hole also.

Since this particle is always pushing away anything around it, it can not gather together with other particles to form the substance. Since it's an end of a despace hole of the Future, the despace of the Future could only be affected or changed by the Future, the anti-gravitational particle is almost impossible to be controlled or affected by the particles or fields of our world.

When an anti-gravitational particle passes through a substance, it will affect the substance and hence be detected with mass, not anti-gravitation.

1.13 Time

There is not a thing named "Time" in the world. Time is only a concept, a calculated value to describe the rate of change between differences. The change of the world is not caused by Time, but the change of world to be described with the concept of Time.

Time is an axis of a 4-dimensional space. It's all the same with how we define the distance. Time is only a reference value with respect to another reference.

1.13.1 Reference of Time

A movie is constructed by numbers of frames. Each frame will be given by a frame mark. When the movie plays, we will see a 2+1 dimensional ($XY+T$) world. There are people and cars moving, sunrise and sunset in the movie. If a man in the movie carries a watch shows the time, and another man with another watch shows its own time. We will find them have self reference time for themselves, and a world reference time with respect to those people who are watching the movie.

No matter how fast nor how slow the movie plays, the self reference time for the men in the movie is always invariant. But for the one who is watching the movie, different playing speed results in different playing time. Because the world reference time is different. The self reference time could never be changed, but the world reference time may change somehow, and the men and anything in the movie will neither notice about it, nor affected by it. For every frame of the movie, the things in the movie are static. While as the movie plays, we see them move.

1.13.2 Chaos

If we project all the frames of a movie on a screen at the same time, we create a 2 dimensional Chaos. A Chaos is a mess without order, but somehow one can find orders within it. The screen we project the movie, it's a Time-Domain for the movie.

As the movie plays, we are seeing an instant view of each frame. It's not possible to go back or go to the future for what we see. What you have already seen, you can't turn to you haven't seen. What you have not yet seen, you can't make you to be already seen.

1.13.3 Time Domain

We are living in a 3D world. However it is almost all the same to a 2D world, except for the depth (or Z axis) does not exist in a 2D world. The-current is like the projection of a movie. Our world is moving towards a direction which we call the-future usually in the Chaos. Our time domain will always be the-current of our world. Everyone has his own self-reference time. The world has its own world-reference time. Different world means different time domain. There could be infinity of time domains. In other words, there could be infinite of different worlds ahead or behind of our world in time dimension.

CHAPTER 2

PHOTON & HALFTON

2.1 Photon and Halfton

Light, which exists in tiny "packets" called photons, exhibits properties of both waves and particles. This property is referred to as the *Wave–Particle Duality*. In some aspects, light could be expressed as a waveform, such as the interference of light wave, the polarization of light wave, and the diffraction of light wave. However, in some other aspects, the light is more than a particle, such as the light pressure, the photoelectric effect, and the reflection of light.

We know that the light is some kind of a waveform. However, the photons are not continuous as a wave though even they are positioned very closely. They are all separated as individual particles. One photon does not affect the other photons before or after themselves. The reason is that a photon is not a waveform, but a particle(s) moves in a course of a waveform.

2.1.1 The Halfton

I hereby present my postulation about the photon. A photon is not "A Particle". A photon is actually composed of two of particles, a *Halfton* and an *Anti-halfton*. The anti-halfton is an anti-particle of the halfton. The phase difference of the time-space vibration between a halfton and an anti-halfton is 180°. However, there is no absolute halfton and anti-halfton, the halfton and the anti-halfton are all the same except for the phase of the time-space vibration. Any two of halftons with phase difference of 180° can form a pair of a halfton and an anti-halfton which forms a photon.

Basically, the halfton and the electron are the same things. They are the same time-space vibration. However, the phase difference of the time-space vibration of a halfton with respect to an electron can vary from 0 to 360°. The halftons are actually part of the electrons. They will resonate with the electrons when they enter the atoms and merge into the electrons. The numbers of the halftons and the anti-halftons merged into the electrons of an atom are equal and balanced with each other. They do not affect the property and charge of the atom.

2.1.2 Emission of a Photon

We can take a halfton and an anti-halfton as an electron and a positron. The electrostatic force between a halfton and an anti-halfton is all the same with which of an electron and a positron (*Please refer to chapter 1, Part 1.6*).

When a halfton and an anti-halfton are attracted by each other and getting closer to each other, they will eventually become to revolve around their mass center. They will accelerate while their revolution orbit is getting smaller. However, with increase of the revolution speed, the revolution orbit will become larger consequently. This will eventually attain to a balance state when their attraction force and the centrifugal force counterbalance with each other.

While the halfton and the anti-halfton are revolving around their mass center, the halfton and the anti-halfton are also vibrating their vicinity space. When their duty cycle of the revolution and the time-space vibration are equal, the halfton and the anti-halfton revolving around their mass center can be displayed as the following picture.

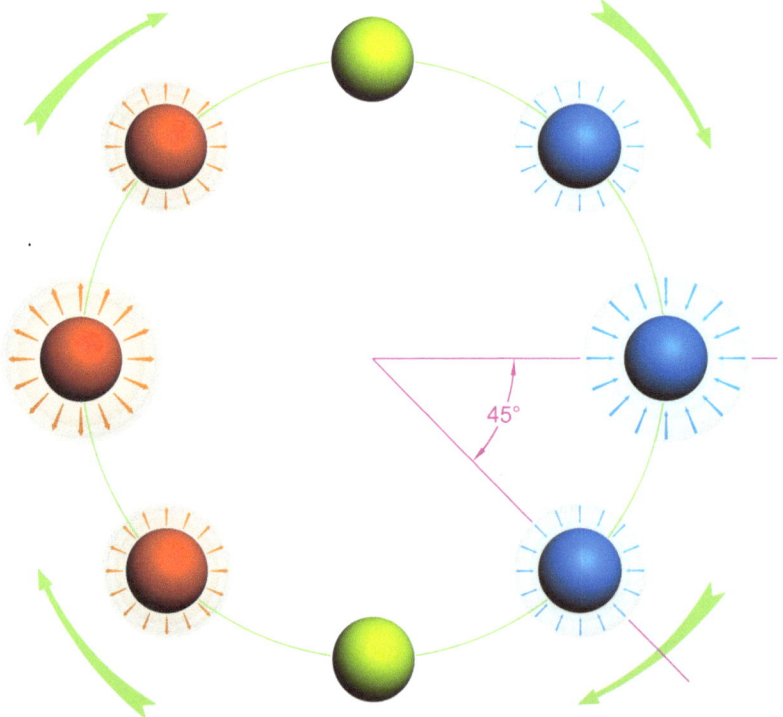

(Fig2.1-1: Revolution of a halfton and an anti-halfton)

For the time-space vibration, when the halfton is at the maximum level of the space shrinkage, the anti-halfton is exactly at the maximum level of the space distension. If we place the halfton at the maximum level of the space shrinkage at the right quadrant position of their revolution orbit, the anti-halfton at the maximum level of the space distension should be at the left quadrant position. Since they are revolving around the mass center and also vibrating the space with the same frequency, the halfton will be replaced by the anti-halfton, and the anti-halfton will be replaced by the halfton for every half of the duty cycle. The particle shrinking the space will always appear at the right side, and the particle distending the space will always appear at the left side.

The mass center will be vibrated by the electrostatic force of the halfton and the anti-halfton back and forth if they do not revolve around. However, along with their revolution around the mass center, the mass center will always be vibrated rightwards in a sine wave oscillation, and vibrates one duty cycle within per half cycle of the revolution of the halfton and the anti-halfton.

Assume the halfton and the anti-halfton are revolving clockwise：

At 0°   The mass center will be accelerated full rightwards. And the photon will start to move rightwards.

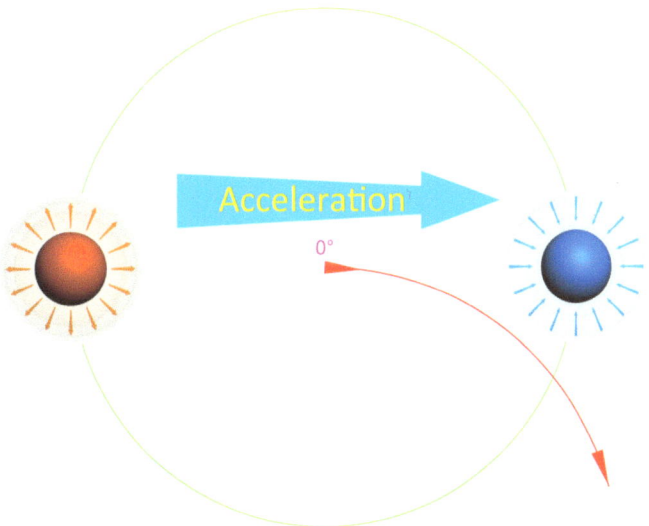

(Fig2.1-2: 0° phase of revolution of halftons)

At 45°   The mass center will be accelerated half rightwards and half down-
          wards. But the acceleration is also decreasing. The photon will move a
          little downwards and keep moving rightwards.

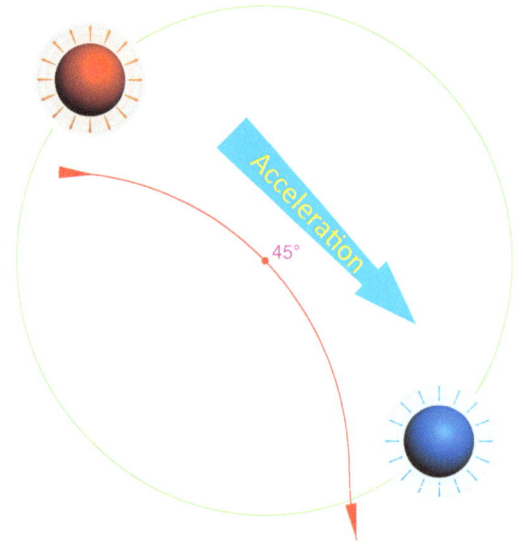

(Fig2.1-3: 45° phase of revolution of halftons)

At 90°   The mass center will no more be accelerated. The photon will keep
          moving downwards and rightwards.

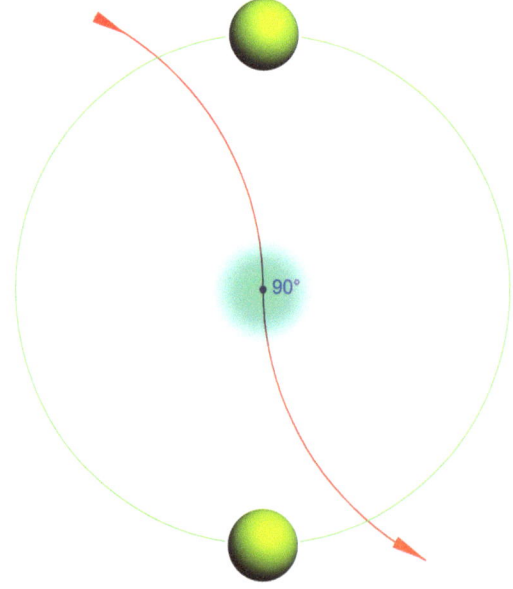

(Fig2.1-4: 90° phase of revolution of halftons)

At 135°   The mass center will start to accelerate upwards and rightwards. The upward acceleration is going to counterbalance the downward acceleration between 0~90°. The photon will keep moving rightwards and decelerate moving downwards.

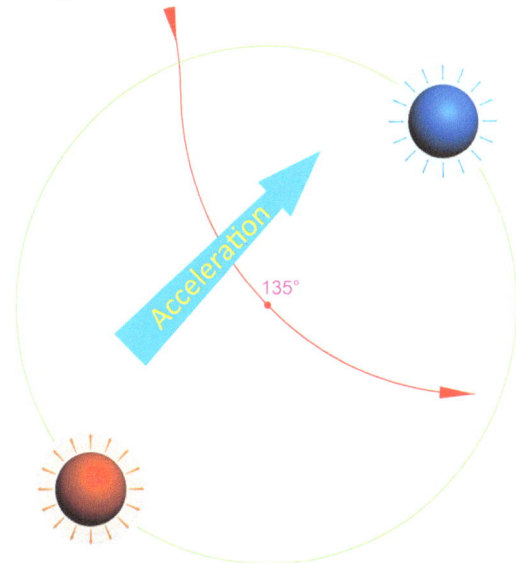

(Fig2.1-5: 135° phase of revolution of halftons)

At 180°   The mass center will be accelerated full rightwards. The downward acceleration has been totally counterbalanced with the upward acceleration. The photon will eventually become to move at its original direction at 0°.

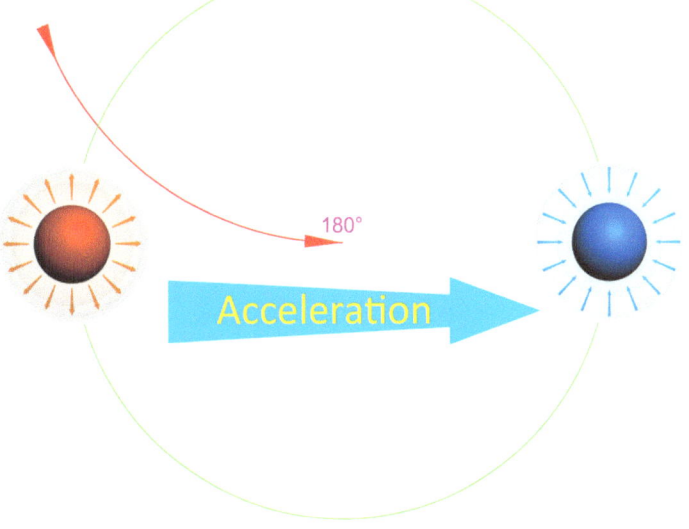

(Fig2.1-6: 180° phase of revolution of halftons)

This oscillation will happen within every half cycle (180°) of the revolution recurrently. The photon will always move rightwards during every cycle, and be accelerated and decelerate vertically. This will cause the photon to emit at an angle of less than 45° in a sine waveform.

As a conclusion, a photon is composed of two particles but moves in a course similar to sine waveform. Since the photon vibrates perpendicular to its propagation direction, thus the waveform of the course of a photon is a 2-D *Transverse Wave*. And the light wave is actually the space vibration at the center of the half-ton pair. A pair of particles moves in a course of a sine waveform.

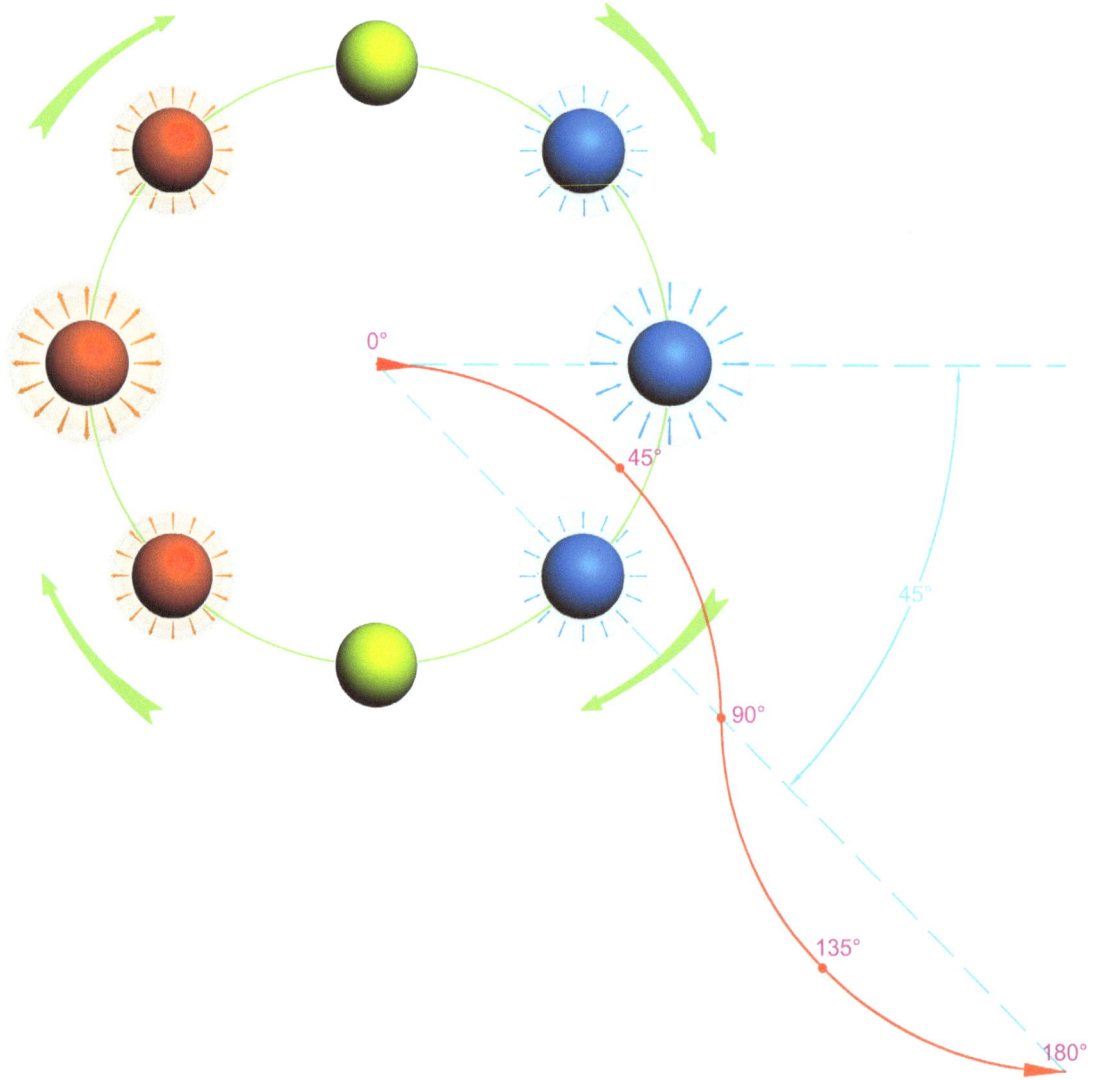

(Fig2.1-7: Moving course of the mass center of a photon)

2.1.3 Halfton and Thermo Energy

The space of the mass center of the photon will be vibrated along with the revolution of the halftons. If the photon emits, the space vibration of its mass center will be a sine waveform. If the photon does not emit but the halftons revolve around an almost fixed center, the space vibration caused by the halftons will vibrate this center as well even the halftons are merged into the electrons.

When a photon composed of halftons collides with the atom, it will be caught by the atom very easily. The halfton and the anti-halfton can be torn apart or be absorbed together in one atom.

The halftons will revolve around the nucleus after being absorbed by the atom and merged into the electrons in the form of the time-space harmonic. This means the halftons are not actually merged for good. The time-space harmonic of the halftons and the electrons can be broken easily and spontaneously. The electrons will gain the energy of the halftons when merged. And hence come into the higher energy state or orbit.

The numbers of the halftons and anti-halftons which have been merged into the electrons will be always balanced. In this case, the electric property or the electric charge of the atom will not be affected.

When the electrons revolve around the nucleus of an atom, the space vibration caused by the electrons will vibrate the nucleus. This will cause the atom to vibrate. The more the energy of the electrons absorb, the greater the atom vibrates generally. This is the so called *Thermo Motion* or *Thermo Vibration*.

When an electron transfers from a higher energy state into a lower energy state, it will break the time-space vibration harmonic and release the halftons. The halftons will emit as a photon if their revolution orbit and their time-space vibration meet with the condition of a photon emission. This is the so called *Thermal Radiation*. The halftons will also be able to be re-absorbed by the electrons possibly.

2.1.4 Cosmic Microwave Background Radiation

The *Cosmic microwave background* (*CMB*) radiation is a form of the electromagnetic radiation filling the universe. With a traditional optical telescope, the space between stars and galaxies (the background) is black. But with a radio telescope, there is a faint background glow, almost exactly the same in all directions, that is not associated with any star, galaxy, or other object. This glow is strongest in the microwave region of the radio spectrum, hence the name cosmic microwave background radiation.

When the space is pouring from the Future into our world, it's like the rain drops falling on the surface of water and causing the water to bounce up and down on the surface. There will always be time-space vibrations caused by the space flow of the universe with different phase. If the phase difference of any two of the time-space vibrations happens to be about 180° and they are close enough, the two of the time-space vibration will form as a pair of a halfton and an anti-halfton as a photon and emit as radiation. The frequency of the photon is based on the frequency of the time-space vibration of the halftons. Since the frequency of the time-space vibration of the halftons is almost constant, the original frequency of the CMB radiation will be almost constant as well.

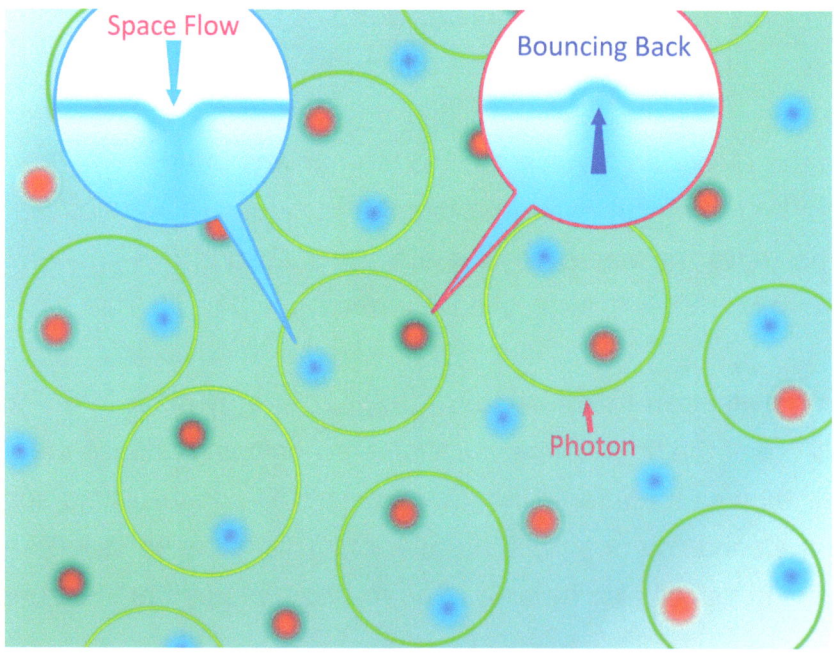

(Fig2.1-8: CMB radiation generated by universe space flow)

2.2 Light Speed and Energy

2.2.1 Light Speed vs. the Revolution of the Halftons

According to the propagation model of a photon demonstrated in *Part 2.1*, we can express the speed of light as below equation:

$$c = K \times r \times \omega \qquad \text{(i)}$$

c is the light speed in vacuum.

K is the constant of displacement of revolution.
       Which means the displacement of each revolution made by the halftons.

r is the mean radius of the revolution orbit of the halftons, which equals to half of the wavelength of the light beam.

ω is the angular speed of the revolution of the halftons

In this case, the light speed is the average speed made by every cycle of the halfton revolution in a revolution orbit of the mean radius of r and at an angular speed of ω. And ω is just equal to the angular speed of the time-space vibration of the halftons, because the halftons vibrates one cycle of the time-space vibration per their revolution around their mass center.

Although the mass center of the halftons is propagating in the space, however, the halftons are still propagating in time dimension actually. Since the speed of light in its medium is constant, this means ω is inversely proportional to r. That is to say, the higher the frequency of the halftons, the smaller the revolution orbit of the halftons. The stronger electrostatic force of the halftons will cause the halftons to get closer while revolving around their mass center. And hence the higher frequency of the time-space vibration of the halfton, the more the energy it contains.

2.2.2 Energy of a Photon

By analyzing the kinetic energy of a photon, the moment of inertia of a photon can be described as below:

$$I = mr^2 \qquad \text{(ii)}$$

Where m is the total effective mass of the halftons of a photon (note that the halftons and the photon do not actually have mass)

Hence the angular kinetic energy of a halfton and an anti-halfton around their mass center equals to K_r:

$$K_r = \frac{1}{2} I \omega^2 \qquad \text{(iii)}$$

$$\rightarrow K_r = \frac{1}{2} mr^2 \omega^2 \quad , \because \ c = \kappa r \omega \implies r\omega = c/\kappa \qquad \text{(iv)}$$

$$\rightarrow K_r = \frac{1}{2} m \left(\frac{c}{\kappa} \right)^2 = \frac{1}{2\kappa^2} mc^2 \qquad \text{(v)}$$

The kinetic energy of a photon with light speed c equals to K_c:

$$K_c = \frac{1}{2} \cdot mc^2 \qquad \text{(vi)}$$

Hence the total energy of a photon contains equals to $E = K_r + K_c$

$$E = K_r + K_c = \frac{1}{2} \cdot mc^2 + \frac{1}{2\kappa^2} \cdot mc^2 \qquad \text{(vii)}$$

$$\rightarrow E = \left(\frac{1}{2} + \frac{1}{2\kappa^2} \right) mc^2 = \left(\frac{\kappa^2 + 1}{2\kappa^2} \right) \cdot mc^2 \qquad \text{(viii)}$$

$$\rightarrow E = \zeta mc^2 \qquad , \quad \zeta = \frac{\kappa^2 + 1}{2\kappa^2} \qquad \text{(ix)}$$

Meanwhile, according to Planck's law:

$$E = h \cdot f \qquad \text{(x)}$$

Where h is the Planck constant, f is the frequency of light.

$$\rightarrow E = \zeta mc^2 = hf \qquad \text{(xi)}$$

$$\Rightarrow m \propto f \qquad \text{(xii)}$$

In this case, since the light speed c of a photon is constant in its medium, thus the energy of a photon concerns only with its effective mass m. In the mean time, according to Planck's law, the energy of a particle contains concerns only with its frequency. In this case, the effective mass of a photon concerns with its frequency.

2.2.3 The Relative Velocity of Photon and Halfton

Since the halftons are carried by the atoms before they emit as photons, though the halftons revolve around the nucleus in almost light speed, however the mean relative velocity of the mass center of the halftons with respect to the nucleus is always zero. Thus, before a photon emits, the mean velocity of the halftons equals to its carrier's velocity.

Before the halfton and the anti-halfton are composed into a photon, they are revolving around the nucleus of their carrier in a speed of about light speed. In this case, at the instant of the formation of a photon, the photon will move in the speed of light in its medium. It won't need the time to accelerate.

This relative velocity of light with respect to the space can be described as below:

Case A : $\vec{c'}=\vec{c}+\vec{v}_m$

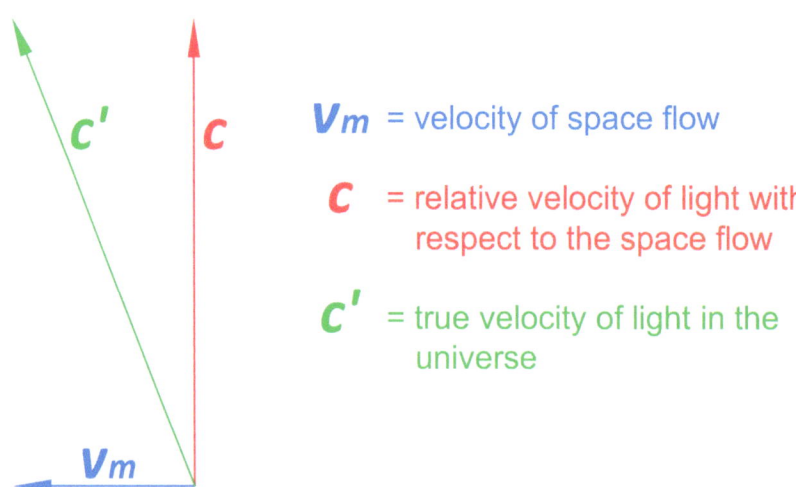

V_m = velocity of space flow

c = relative velocity of light with respect to the space flow

c' = true velocity of light in the universe

(Fig2.2-1: The velocity of light with respect to space and universe)

In this case, the relative velocity of the light source with respect to the space (the medium of light) is zero.

However, if the carrier of the halftons (the source of light) has a relative velocity with respect to the space (the medium of light), which means the halftons will also have the same relative velocity with respect to the space before they emit as photons. In this case, at the instant of the formation of a photon, the emitted photon will also have the same component of the relative velocity of its original carrier with respect to the space. However, the relative velocity of light with respect to its medium will be limited to the light speed in that medium regardless of the velocity of its source. This can be described as below:

Case B : $\vec{c} = \vec{\hat{c}} + \vec{v}_s$

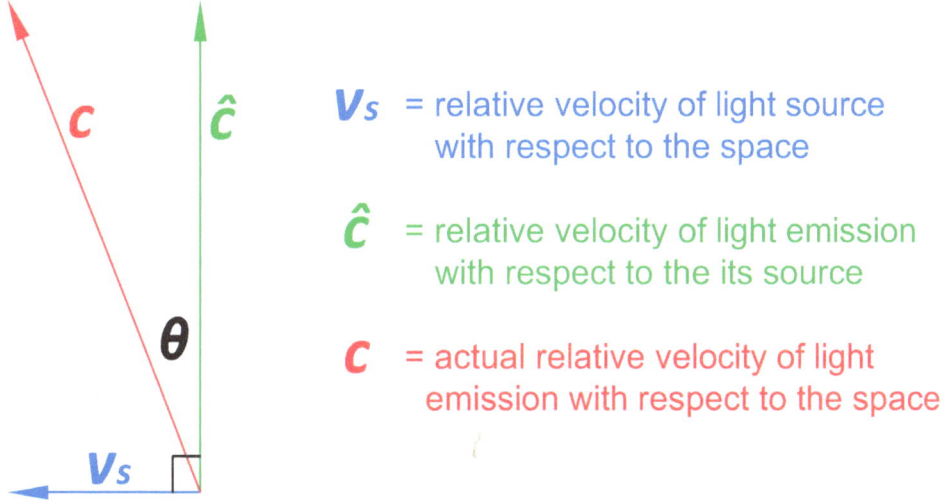

V_s = relative velocity of light source with respect to the space

$\hat{c}$ = relative velocity of light emission with respect to the its source

C = actual relative velocity of light emission with respect to the space

(Fig2.2-2: The relative velocity of light with respect to its source)

Note that above picture is shown as the space is considered as stationary. That is to say, v_m is considered as 0 to simplify our demonstration. This actually affects the direction of the light propagation if v_s is not parallel to c since the direction of light with respect to its source will always be invariant. And the velocity of the light source will become a part of the velocity of light. In this case, the relative velocity of light with respect to its source will vary in accordance with the velocity of its source.

2.2.4 Doppler Effect of Light

The energy of the halftons should conserve all the time while they are inside the atoms or emit as photons. In this case, the energy of the emitting photon contains will equal to the energy of the halftons plus the kinetic energy given by the light source. This can be shown as below:

$$\vec{E}_e = \vec{E}_c + \vec{E}_s$$

E_e = Energy of the emitting photon

E_s = Energy given by the light source

E_c = Original energy of the photon

(Fig2.2-3: The energy of the emitting photon)

Since the light speed in the same medium is constant, thus the higher the speed of the light source, the higher the frequency of the emitting photon. This can be described as below:

$$E_0 : E_e = hf_0 : hf_e \qquad \text{(i)}$$

$$\Rightarrow E_0 : E_e = f_0 : f_e \qquad \text{(ii)}$$

Where $E_0 = E_c + E_{s0}$, $E_e = E_c + E_s$

$$\rightarrow \left(E_c + E_{s0} \right) : \left(E_c + E_s \right) = f_0 : f_e \qquad \text{(iii)}$$

Let $E_{s0} = 0$ for the light source is stationary with respect to the space.

$$\rightarrow E_c : \left(E_c + E_s \right) = f_0 : f_e \qquad \text{(iv)}$$

$$\Rightarrow f_e = \frac{E_c + E_s}{E_c} \cdot f_0 \qquad \text{(v)}$$

If there is a receiver or a detector at a relative velocity with respect to the space which will receive and detect the photon, the photon will have to contain more energy equals to the kinetic energy of the receiver to be absorbed. This can be described as below:

$$E_0 : E_R = f_0 : f_r \qquad \text{(vi)}$$

Where $E_0 = E_c + E_r$, $E_R = E_c$

$$\rightarrow \left(E_c + E_r \right) : E_c = f_0 : f_r \qquad \text{(vii)}$$

$$\Rightarrow f_r = \frac{E_c}{E_c + E_r} \cdot f_0 \qquad \text{(viii)}$$

In this case, the relation of the frequency of the photon between the emitter and the receiver can be derived from the superposition of the two effects:

$$f_e = \frac{E_c + E_e}{E_c + E_r} \cdot f_r \qquad\qquad \text{(ix)}$$

For the photon in the same reference frame, if there is no relative velocity between the light source and the receiver, there will not be frequency change since $E_e = E_r$.

However, this is different from the Doppler Effect since the photons do not affect with each other though even their results are similar. In other words, this is not the Doppler Effect.

Thus, we can conclude as following :

1) *The relative velocity of light with respect to its medium is constant.*
2) *The relative velocity of the light source with respect to the space will affect the direction and the frequency of the light emission.*

2.2.5 Light Speed in Different Medium

The pressure of the space (**POS**) is almost constant in the universe. However, at the place where substance highly concentrated, the **POS** will be reduced by the space diminution of the despace holes. In this case, in an area with more density of the substance, the pressure of the space will become lower comparatively. The more density of the substance is defined as the *Less Optically Dense Medium*, and the less density of the substance is defined as the *More Optically Dense Medium*.

Within the less optically dense medium where the pressure of the space is lower, the space vibration caused by the halftons will take more effect to the space. And within the more optically dense medium where the pressure of the space is higher, the space vibration caused by the halftons will take less effect to the space comparatively.

As the effect of the space vibration of a particle becomes more, the weight effect which it generates will increase accordingly (see *Part 1.5.3*). Thus the effective mass of the halftons will increase consequently. Note that though even the effect of the space vibration and the effective mass of the halfton may change, the halfton itself does not change actually.

As the effect of the space vibration of the halftons becomes more, the attraction force between the halftons will become stronger consequently. This will reduce the radius of the revolution orbit of the halftons around their mass center.

$$\because c(\downarrow) = \kappa \times r(\downarrow) \times \omega$$

This will reduce the light speed in that medium consequently.

Though the effective mass of the halftons will change, however, the frequency of the halfton will not change. It only changes the relative effective mass of the halftons with respect to the local **POS**.

As the effective mass and the light speed are concerned for the energy of the photon, the effective mass of the photon increases, on the other hand, the light speed reduces. As a result, the energy of the photon will remain the same.

$$\because E = \zeta mc^2$$

From another point of view, the frequency of light remains constant, the energy of light will remain constant as well. But the light speed will reduce as the radius of the revolution orbit of the halftons will reduce in that medium.

$$\because E = hf$$

2.2.6 Refraction

Refraction happens when a light beam passes from one medium into another. The light beam will change its direction due to change of light speed in different optically dense medium.

Since change of light speed is actually the change of its wavelength which is the diameter of the revolution orbit of the halftons, this can be taken as change of size of a photon.

The halftons may revolves on any 2D plan and become as a wave packet as a 3D ball. When the photons travel from one medium into another, the transformation of a photon wave packet can be demonstrated as below picture:

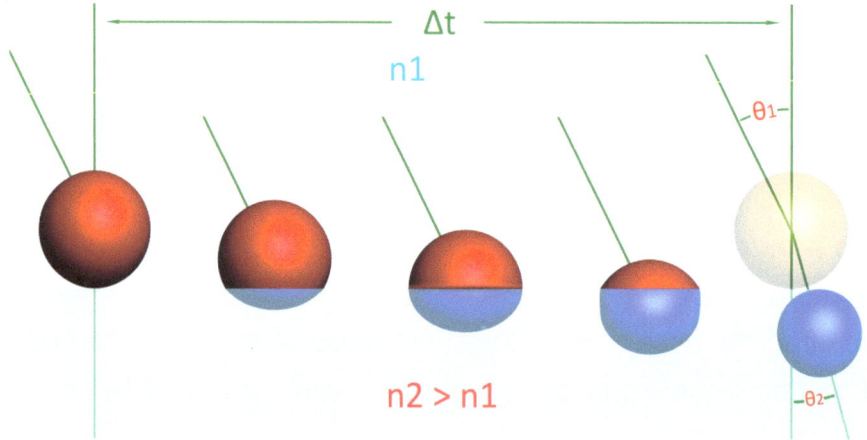

(Fig2.2-3: Direction change of a photon wave packet)

Since the photon is partially deformed, the mass center of the upper and lower part of the photon will move in different speed and cause the photon to rotate and hence cause the photon to change its direction.

For all the photons with different frequency have the same speed in the same medium. Thus, for the smaller photon (with higher frequency), it takes less time for it to move from one medium into another one comparatively. In this case, the speed difference of the upper and lower part of the photon will become greater and hence causes the rotation of the direction of the photon to become grater.

2.2.7 Cherenkov Radiation and Faster-Than-Light

Cherenkov Radiation is the electromagnetic radiation emitted when a charged particle (such as an electron) moves at a speed greater than the speed of light in that medium. For example, the speed of the propagation of light wave in water is only $0.75c$. Substance can be accelerated beyond this speed during nuclear reactions and in particle accelerators. Cherenkov radiation results when a charged particle, most commonly an electron, travels through a dielectric (electrically insulating) medium with a speed greater than that at which light propagates in the same medium.

When an electron is moving faster than light in that medium, the space rebounds caused by the space vibration of the electron will become as a spring along with the path of the moving electron, and hence be left behind of the electron and becoming into halftons. As each cycle of the space vibration of the electron will have one maximum positive rebound and one maximum negative rebound which will become into one halfton and one anti-halfton as a pair of a photon.

The distance between the halftons is proportional to the moving speed of the electron. Since the speed of the halfton is fixed to the light speed in that medium, the frequency of the halftons will be inversely proportional to the distance between the halftons. The faster the electron moves, the longer distance between each halfton. As a result, the frequency of the photons will decrease due to increase of the speed of the electron.

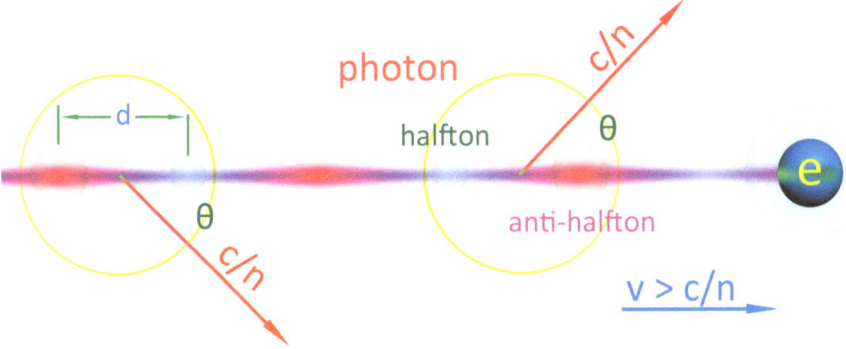

(Fig2.2-4: Cherenkov radiation of an electron)

Since the speed of all the photons in the same medium is c/n, where n is the refractive index of that medium, thus the faster the electron moves, the lower the frequency of the Cherenkov radiation. Meanwhile, with increasing of the moving speed of the electron, the number of the created photons will become more and hence increase the density of the radiation.

Though the original speed of the halftons is v which is faster than light in that medium, however, the speed of the photons are limited to c/n. In this case, the speed of the photons can be taken as a component of v. And the emission angle θ of the photons would comply with the following equation:

$$\cos\theta = c\Big/nv$$

2.3 Interference

The interference of the light waves looks very similar to the general mechanical waves. However, it's actually different, because the light wave is not a wave.

2.3.1 Propagation of Transverse Wave
For a transverse wave, it is the whole waveform which moves in the medium as demonstrated below.

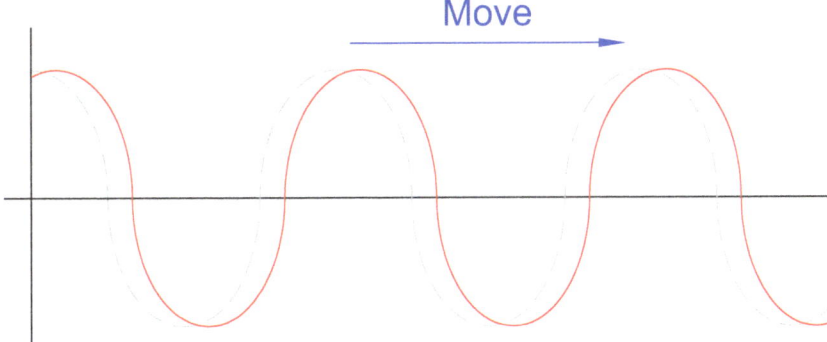

(Fig2.3-1: How a transverse wave moves)

For the photons, it is the individual photon which moves in the course as demonstrated below.

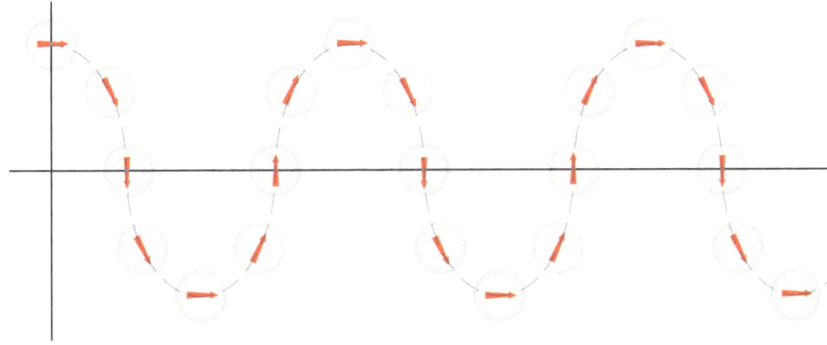

(Fig2.3-2: How the photons move)

2.3.2 Interference

For the transverse waves with constant wave length and different phases, there will be construction and destruction of the waveforms. If the phase difference of all the waveforms are even and smooth, the harmonics may become a total destruction of waves as a straight line. Or otherwise, the harmonics may become as a new waveform with the same wave length and frequency but different amplitude.

For the mechanical transverse waves, harmonics of the waves move in the same way with one single waveform. However, the light wave will become as a fixed pattern of alternate brightness and darkness as so called interference fringes.

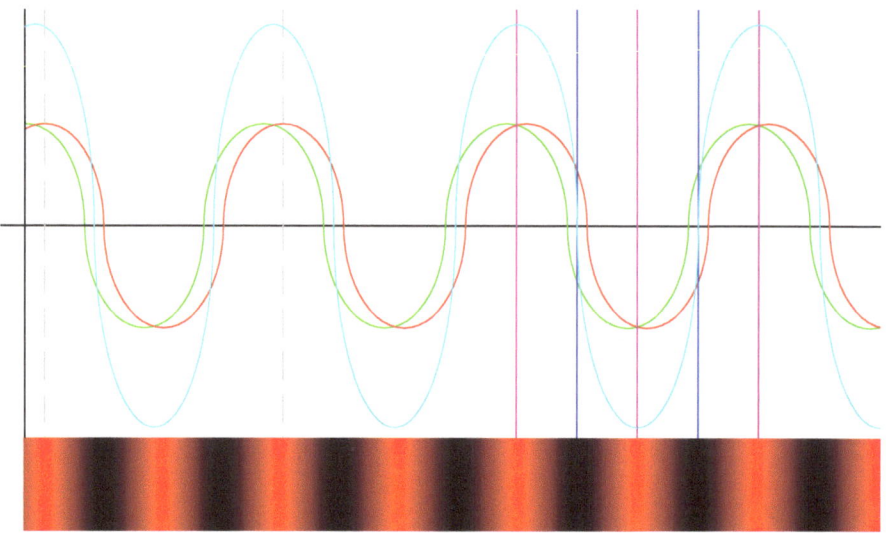

(Fig2.3-3: Interference of waves and lateral view of light pattern)

The interference of photons happens when multiple photons collide with the same atom at the same time. The space vibration of the halftons which will vibrate the nucleus will interfere with one another. The space vibrations to the nucleus with the same direction can be multiplied as constructive interference (the bright region); the space vibration to the nucleus with opposite direction will be diminished as destructive interference (the dark region).

2.3.3 Diffraction

Diffraction of light happens when a light beam passes through a tiny slit. Though the slit may be a very thin slice, however the thickness of the slice is long enough for the photons to scatter over the wall on the path. After the photons pass through the slit, they will scatter at any angle and interfere with one another as well.

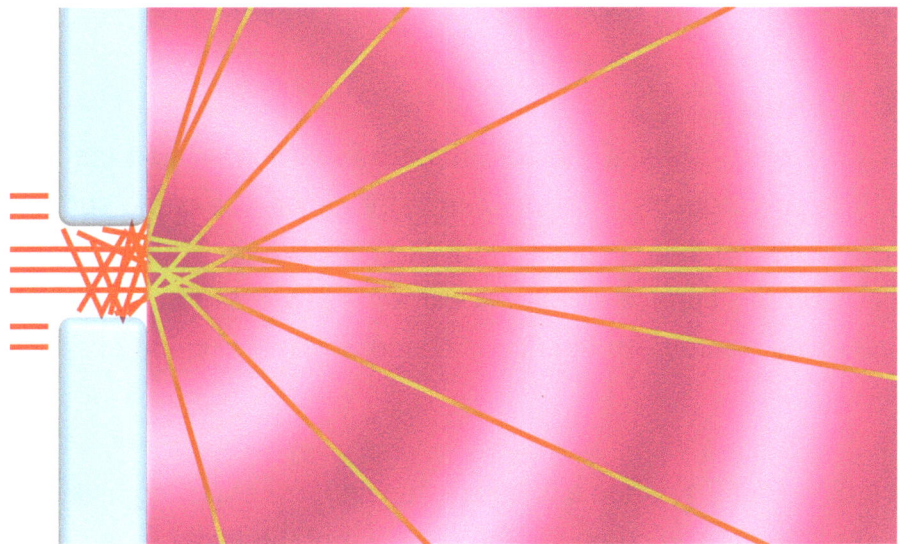

(Fig2.3-4: Diffraction of light wave)

For the single side slit, the curvature on the thickness of the slit and the air may also cause the photons to scatter and cause single side diffraction.

2.4 Collision of Photon with Matter

There are three kinds of situation for a photon collides with matter, such as *Reflection*, *Photoelectric Effect*, and *Thomson and Compton Scattering* resulting from the difference of the wave length of the light wave.

2.4.1 Reflection

Since the electrons and the halftons are revolving around the nucleuses all the time, they are probably to be pushed to the outer side or the surface of matter when they are distending, especially for the conductors. Thus, the electrons and the halftons on the surface of matter will form a wall of force field that pushing away everything nearby. And hence produce a mirror surface to reflect the light waves.

According *Part 2.1*, when a photon is moving, the shrinking halfton will always be in front of the distending one in the direction of the light propagation. If the radius of the photon is large enough which means the frequency of the photon is comparative low, the wall of the force field will looks tighter and the moving direction of the halftons will become more transverse comparatively with respect to the wall of the force field. Thus the halfton will not be able to pass through it.

In this case, the vertical component of the speed of the photon which perpendicular to the surface of the reflector will be decelerated to zero speed, and then re-accelerate to its original speed in reverse direction. In this case, the leading halfton which will reach the wall will be decelerated first, but the following halfton will keep approaching. The leading halfton will move to a limit distance from the wall before it's going to distending. When the shrinking halfton is becoming into distending, the distending halfton is becoming into shrinking, and the photon will become to move forward again. However, the positions of the halftons are swapped because the shrinking halfton is now the other one which is at the outer side of the wall. This will reverse the direction of the speed component of the photon which is perpendicular to the wall of the force field.

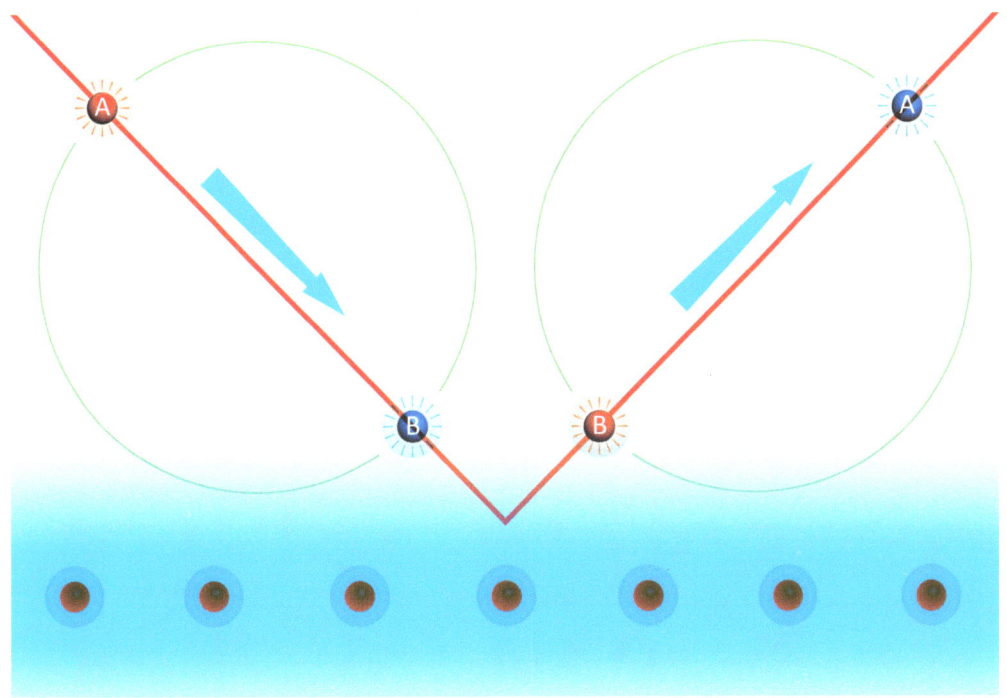

(Fig2.4-1: Reflection of a photon)

Thus, the reflection of the light beam can be simplified as elastic collision of the photon with matter. Since the photon will be decelerated by the matter, the matter will be affected by the weight effect of the photon. This will cause the light pressure on the substance. Note that though even the photon will be decelerated, however, the halftons are still revolving around their mass center. The photon will resume to light speed in the medium immediately after leaving away from the wall.

2.4.2 Photoelectric Effect

If the radius of the photon is small enough which means the frequency of the photon is comparative higher, the wall of force field will become loose. Thus the halfton will have the chance to pass through the wall and get into the matter.

When the halfton is getting close to the matter, it's shrinking at that moment. After it passes the wall of the force field, it will begin to distend. Thus the halfton which is still outside of the wall will be cut off, and the one inside the matter will unite with another halfton which is already inside the matter to form a new pair.

The halfton which is cut off outside of matter will then drift in the air until it is absorbed by the matter. The halfton is almost an electron when it's an individual one. The halftons moving inside a conductor will also cause the similar electric current caused by moving of the electrons.

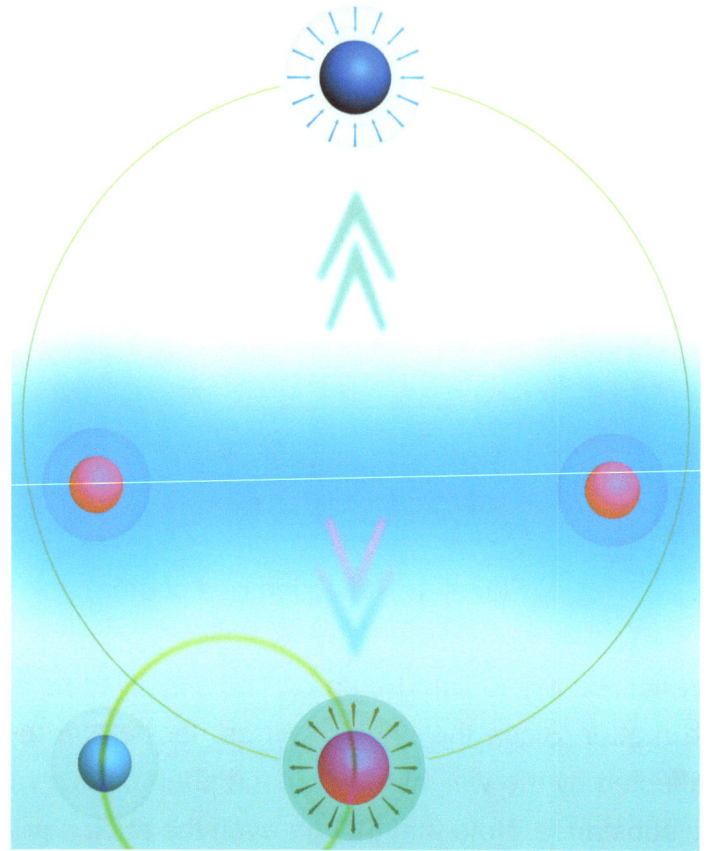

(Fig2.4-2: Photoelectric effect of a photon)

The numbers of the halftons and the anti-halftons inside of the matter are balanced normally. They would form as pairs of halfton and anti-halfton inside the atoms as well. When a halfton enters a conductor and unite with another halfton, this means the original pair of halfton and anti-halfton will be torn apart. And the halfton being torn apart will unite with another anti-halfton inside the matter as well, and hence cause the halftons to move inside the matter as a current.

This is the so called *Photoelectric Effect*. However, what escape from matter are not the electrons but the halftons though they are almost the same things.

2.4.3 Thomson Scattering and Compton Scattering

If the radius of the photon is very small which means the frequency of the photon is very high, the wall of force field will become looser. Thus both the half-ton and the anti-halfton will have the chance to pass through the wall and get into the matter or even pass through the matter.

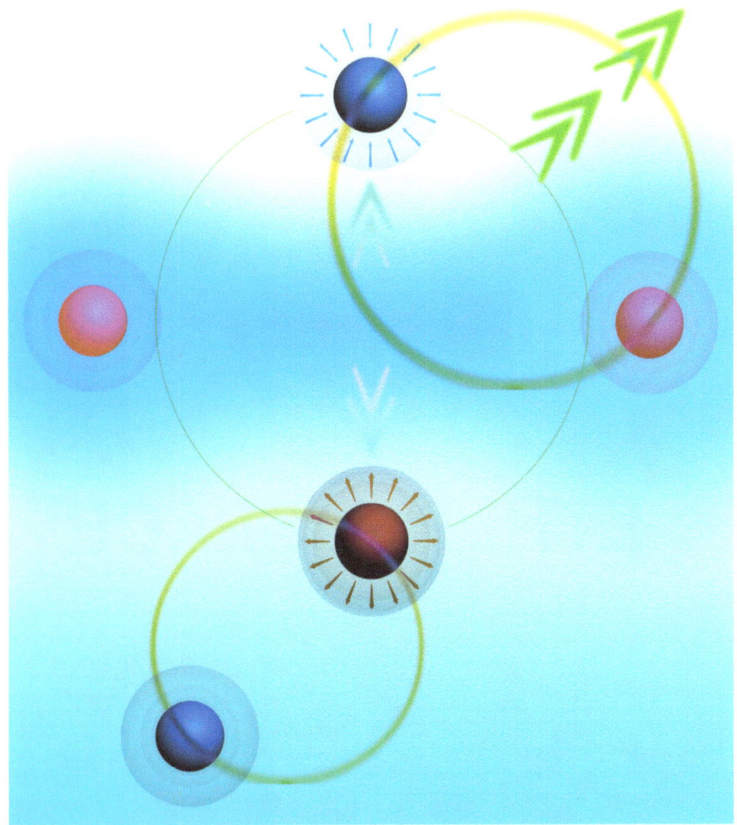

(Fig2.4-3: Scattering of a photon)

When the photon is passing through the wall, the forward halfton may unite with the halfton inside the matter, and the latter halfton which is still making its way passing through the wall may unite with the halfton on the wall and emit as a new photon. The frequency or the direction of velocity of the halfton on the wall will change the frequency of the new photon, and hence cause the *Compton Shift*.

CHAPTER 3

ETHER & SPACE

3.1 Ether

In the late 19th century, physicians came to find out that "Light" is a kind of waveform. However, it normally requires a medium for the wave propagation. Such as audible sound propagates via air and water ripple propagates via water. Since light can travel through vacuum, it was assumed that the vacuum must contain the medium of light. Physicians postulated that there is something named "Ether" exists everywhere in the universe. Thus light and electromagnetic wave can propagate via it. And so does the other action of distance.

3.1.1 Finding of Ether

The *Ether hypothesis* is once a very important theory in Physics. Since Ether is so important, then how can we prove its existence? According to the astronomical phenomenon of *stellar aberration*, it was deemed that the Ether must be stationary in the universe. When Earth moves in the stationary Ether, there must be a relative velocity between Earth and Ether.

In order to measure the relative velocity between Earth and Ether so as to prove the existence of Ether, in year 1881 to 1884, Albert Michelson and Edward Morley cooperated to design the famous *Michelson–Morley experiment* according to the interference principle of light wave and the stellar aberration hypothesis. Since light propagates via Ether as a medium, if the relative velocity of Earth with respect to Ether is changed, the relative light velocity with respect to Earth will change simultaneously.

However, the experiment result proved that the light speed in all direction does not have any difference. Which means the 30km/s revolution velocity of Earth around the sun doesn't affect the relative velocity between an object and light. This contradicts the stellar aberration hypothesis. This experiment did not find the evidence of Ether, but indirectly overruled the Ether hypothesis instead. However, can an experiment contradict with its own basic assumption still be valid? There must be something wrong which beyond our general acknowledge.

3.2 Stellar Aberration

The stellar aberration (also referred to as astronomical aberration or aberration of light) is an astronomical phenomenon which produces an apparent motion of stars about their real locations. It was developed and explained by the third Astronomer Royal, James Bradley, in about September 1728 and his theory was presented to the Royal Society a year later.

3.2.1 Bradley's Discovery of Stellar Aberration

James Bradley found that the observation angle of stars has an annual aberration. And change of the observation angle of a star depends solely upon the transverse component of the observer's velocity with respect to the vector of the incoming light beam velocity. In the case of an observer on Earth, the direction of the observer's velocity varies during the year as Earth revolves around the Sun, and this in turn causes the apparent position of the star to vary periodically over the revolution course of a year. The maximum amount of the aberrational displacement of a star is approximately 20 arc-seconds in right ascension or declination.

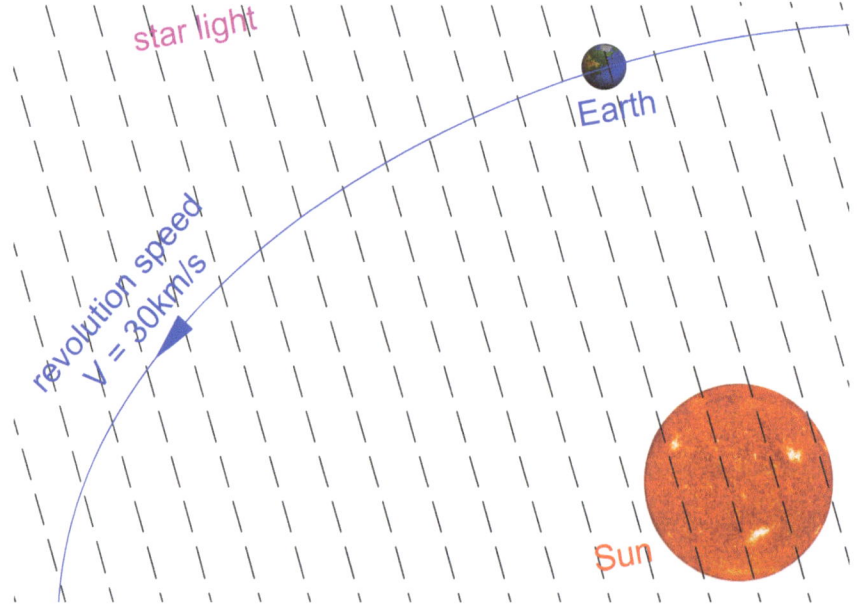

(Fig3.2-1: Starlight passes through the Solar System)

The basic postulation of stellar aberration is that the light beam emitted from a star of an extreme distance can be taken as many parallel lines when it reaches our Solar System. Meanwhile, this theory is also based on the *principle of superposition of velocity*. The transverse component of velocity of an object with respect to the light beam and the velocity of light beam can be combined into a new one. And hence affect the apparent inclination angle of the light beam.

Since Earth revolves around the sun in a velocity of 30km/s, whenever a light beam reaches the ground surface on Earth without dragged by Earth or Ether, there will be an extra transverse component of revolution velocity with respect to the light beam. If we take Earth as static, the light beam will have an extra reversed transverse component of velocity with respect to the observer. Hence if the light beam is originally perpendicular to ground, it will tilt a small angle because there is now a transverse component of velocity.

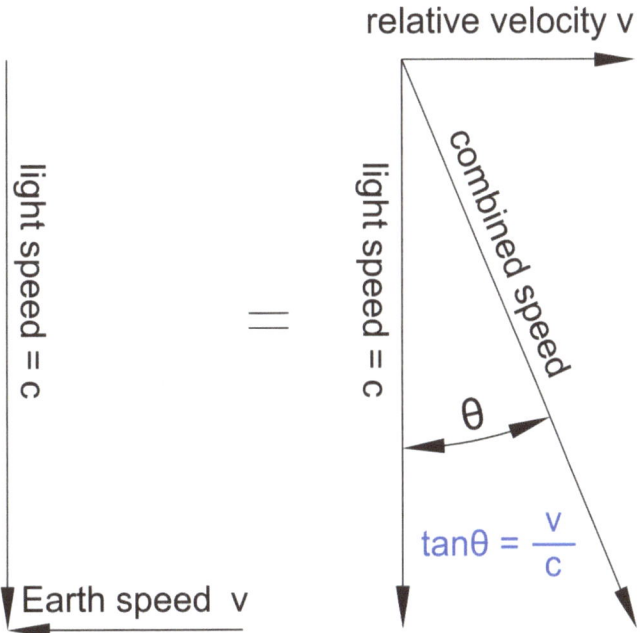

(Fig3.2-2: Aberration caused by observer's transverse velocity)

It's similar to driving a car in the rain. The rain falls from the sky vertically if there's no wind. While as the car moves forward in a speed, the rain will start to decline even if there is still no wind. You will see the rain hitting your windshield in an angle. The higher the speed of your car, the more the decline angle of the rain.

For the observer at the zenith, the light beam is originally tilted with an angle for degrees, but the direction of Earth revolution velocity will change for 360° per year, and hence the transverse component of velocity with respect to the light beam will change from 0 to 30km/s recurrently. The largest ascension or declination of apparent view angle θ can be described as follow.

$$\tan\theta = \frac{v}{c} \approx 10^{-4}$$

$$\rightarrow \theta = \arctan\left(10^{-4}\right) = 5.73 \times 10^{-3} = 20.5" \text{ (arc-second)}$$

The maximum difference in view angle of an observer on Earth over its revolution course of a year will be about $40"$.

3.3 The Michelson–Morley Experiment

The Michelson–Morley experiment, one of the most important and famous experiments in the history of physics, was performed in 1887 by Albert Michelson and Edward Morley at what is now Case Western Reserve University.

If light propagates via Ether, change of relative velocity between Earth and Ether will hence change the light speed. The main purpose of the Michelson–Morley experiment is to detect the change of the light speed so as to detect change of the relative velocity between Earth and Ether.

3.3.1 Michelson Interferometer

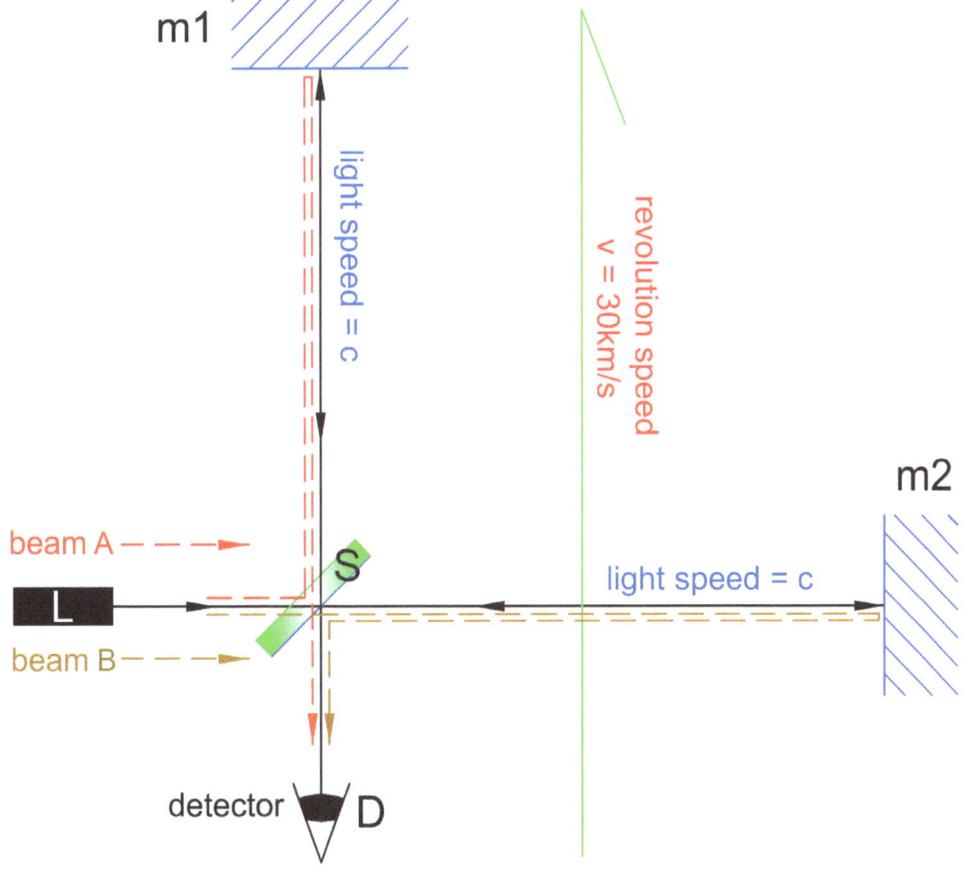

(Fig3.3-1: Michelson interferometer)

Above is the basic construction of the Michelson interferometer. To simplify the explanation, all the lines in above picture are perpendicular to one another. He-Ne laser emits from the light source (denoted as L) to a half-silvered mirror (denoted as S) that was used to split it into two beams travelling at right angles to one another. The light **beam-A** travelled out from S to $m1$, and then was reflected back and pass through S to the detector (denoted as D). The light **beam-B** travelled out from S to $m2$, and was reflected back to S, then travelled to D. The He-Ne laser was split into two beams and then recombined together, producing a pattern of constructive and destructive interference based on the spent time to travel their paths.

If Earth and everything on Earth is moving in a stationary Ether medium, the length of the two paths for the beams reflected back-and-forth perpendicular and parallel to the flow of Ether would be different. The result would be a delay in one of the light beams that could be detected when the beams were recombined through interference. Any slight change in the length of their paths would then be observed as a shift in the positions of the interference fringes.

3.3.2 Laser Beam

A laser beam will always interfere itself on all its way out which will make it to be similar to the below pattern of interference fringes (see Fig3.3-2). As the wavelength of a He-Ne laser is only about 640nm, you would not be able to see these fringes. Besides, you can't see a laser lateral view normally. We see the light because the light hits our eyeball.

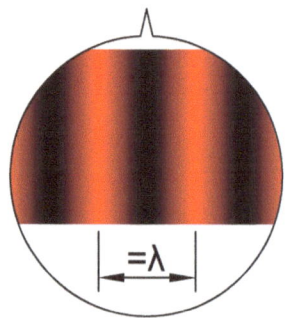

(Fig3.3-2: Lateral view of a laser beam)

The half-silvered mirror placed 45° in above picture (Fig3.3-1) is to slice the laser beam obliquely so as to reflect its lateral view to the mirrors or the detector. Then the observer will be able to see the lateral view of the laser beam.

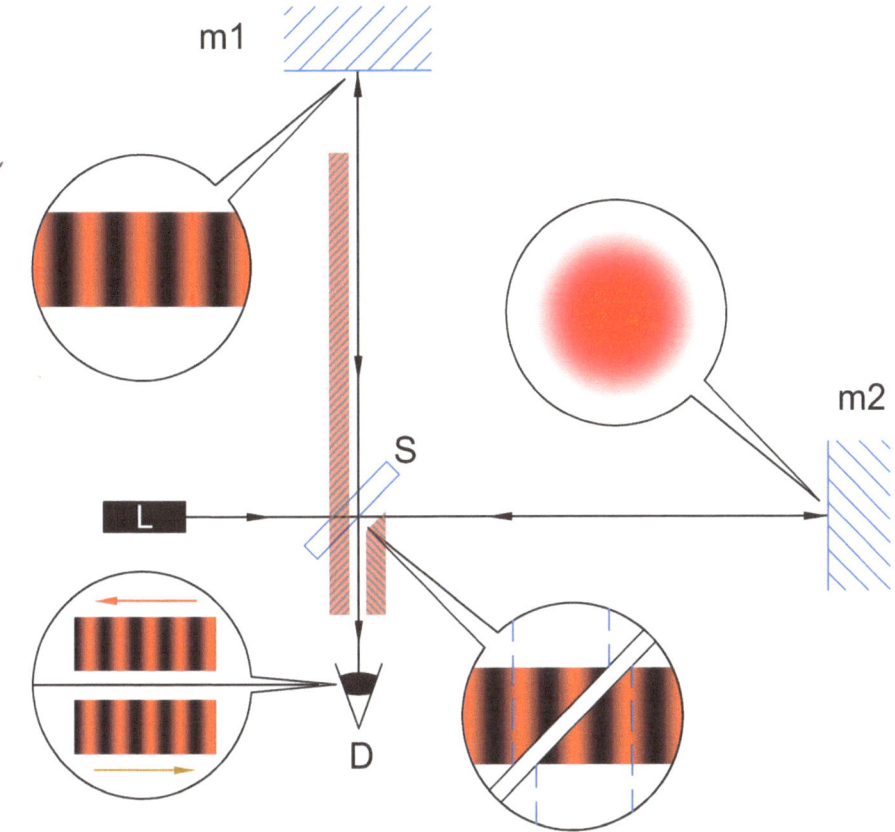

(Fig3.3-3: Laser patterns on interferometer)

3.3.3 Path Difference of Light Beams on Interferometer

Below we displayed two of the light beams in straight lines for easy analysis. The arm length of S to $m1$ and S to $m2$ both equals to L. (Note! The two paths $\overline{LS}$ and $\overline{SD}$ were omitted because they are the same for both light beams):

A) : The path for **beam-A** : $S \rightarrow m1 \rightarrow S$

When **beam-A** left from S to $m1$, $m1$ is also moving away from S at the same time; when **beam-A** reached $m1$, $m1$ has moved for a distance of d_1. So actually, **beam-A** has travelled a distance of ($L + d_1$) for this path.

When light beam was reflected from $m1$ back to S, S is also moving towards $m1$ at the same time; when **beam-A** reached S, S has already moved for a distance of d_2. So actually, **beam-A** has travelled a distance of $(L - d_2)$ for this path.

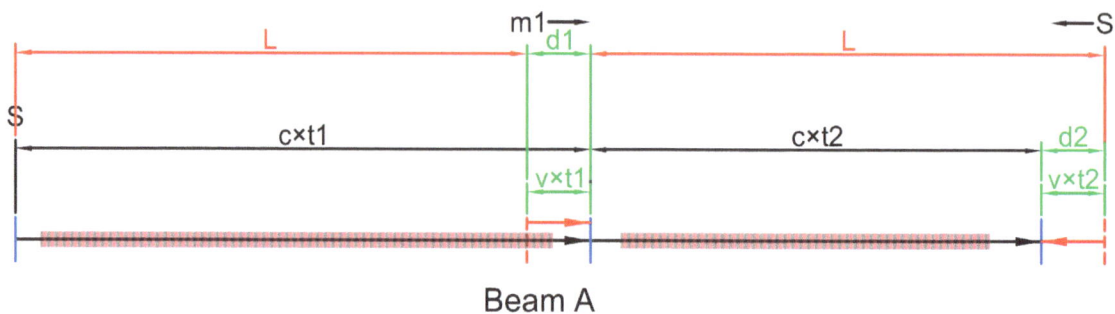

Beam A

(Fig3.3-4: The path of **beam-A**)

The total path length of **beam-A** travelled through $S \rightarrow m1 \rightarrow S$ equals to :

$$(L+d_1)+(L-d_2)=2L+(d_1-d_2) \qquad \text{(i)}$$

Meanwhile, we can calculate the time for **beam-A** to travel from S to $m1$ and the time to travel from $m1$ back to S as below:

$$\begin{cases} L+d_1=c{\cdot}t_1 \\ L-d_2=c{\cdot}t_2 \end{cases} \qquad \text{(ii)}$$

The displacement of $m1$ and S can also be calculated as below:

$$\begin{cases} d_1=v{\cdot}t_1 \\ d_2=v{\cdot}t_2 \end{cases} \Rightarrow \begin{cases} t_1=d_1/v \\ t_2=d_2/v \end{cases} \qquad \text{(iii)}$$

And hence the parameters t_1 and t_2 in equation (ii) can be replaced as below:

$$\Rightarrow \begin{cases} L+d_1=c\cdot(d_1/v) \\ L-d_2=c\cdot(d_2/v) \end{cases} \qquad \text{(iv)}$$

$$\Rightarrow \begin{cases} L=\dfrac{c}{v}\cdot d_1 - d_1 = \dfrac{c-v}{v}\cdot d_1 \\[4mm] L=\dfrac{c}{v}\cdot d_2 + d_2 = \dfrac{c+v}{v}\cdot d_2 \end{cases} \qquad \text{(v)}$$

$$\Rightarrow \begin{cases} d_1=\dfrac{v\cdot L}{c-v} \\[4mm] d_2=\dfrac{v\cdot L}{c+v} \end{cases} \qquad \text{(vi)}$$

$$\Rightarrow d_1-d_2 = \frac{v\cdot L}{c-v} - \frac{v\cdot L}{c+v} = \frac{2v^2 L}{c^2-v^2} \qquad \text{(vii)}$$

Hence the equation (i) can be rewrote as below:

$$\Rightarrow 2L+(d_1-d_2)=2L+\frac{2v^2 L}{c^2-v^2} \qquad \text{(viii)}$$

B) : The path for **beam-B** : $S \rightarrow m2 \rightarrow S$

The path length of **beam-B** travelled through $S \rightarrow m2 \rightarrow S$ equals to :

$$2L \cdot \sec\theta \qquad \text{(i)}$$

$$\sec\theta = \frac{c}{\sqrt{c^2 - v^2}} \qquad \text{(ii)}$$

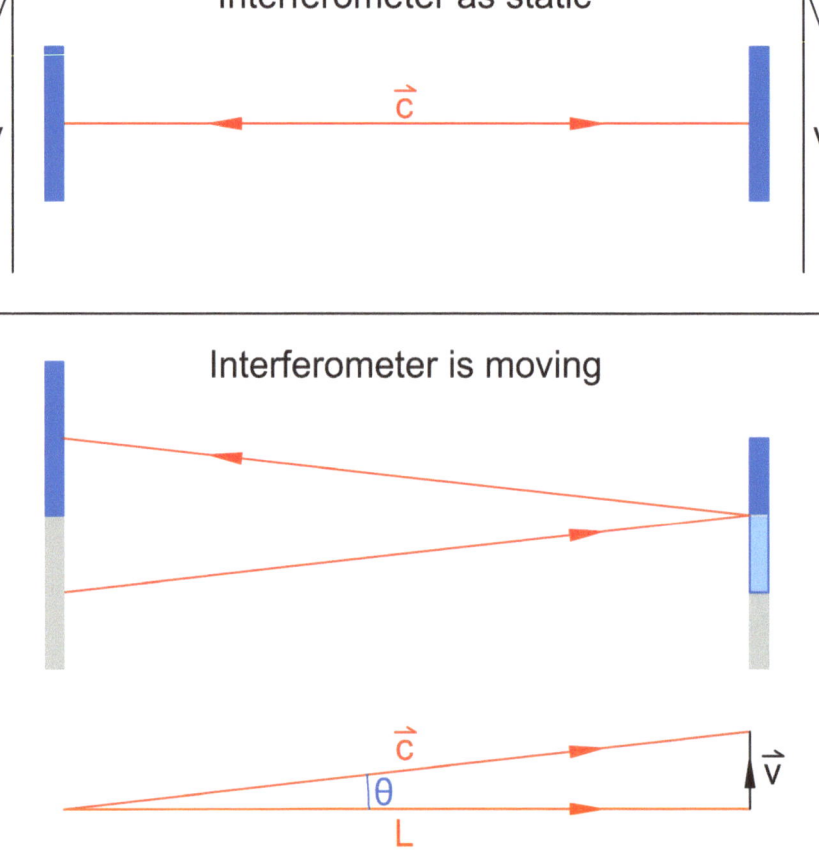

(Fig3.3-5: The path of **beam-B**)

After the whole interferometer was rotated for 90° CCW, The total path length of **beam-A** will become $2L \times \sec\theta$, and the total path length of **beam-B** will become $2L + (d_1 - d_2)$.

In this case, the path difference P_d between the light beams can be calculated as below:

$$P_d = 2\{2L \cdot \sec\theta - (2L + (d1 - d2))\} \tag{i}$$

$$\rightarrow P_d = \{4L \cdot \sec\theta - 4L - 2(d1 - d2)\} \tag{ii}$$

$$\rightarrow P_d = \left\{ 4L(\sec\theta - 1) - 2\left(\frac{2v^2 L}{c^2 - v^2}\right) \right\} \tag{iii}$$

$$\rightarrow P_d = \left\{ 4L\left(\frac{c}{\sqrt{c^2 - v^2}} - 1\right) - 2\left(\frac{2v^2 L}{c^2 - v^2}\right) \right\} \tag{iv}$$

$$\rightarrow P_d = \left\{ 4L\left(\frac{c}{\sqrt{c^2 - v^2}} - 1\right) - 4L\left(\frac{v^2}{c^2 - v^2}\right) \right\} \tag{v}$$

$$\rightarrow P_d = 4L\left\{ \frac{c}{\sqrt{c^2 - v^2}} - 1 - \frac{v^2}{c^2 - v^2} \right\} \tag{vi}$$

$$\rightarrow P_d = 4L \left\{ \frac{c\sqrt{c^2 - v^2} - \left(c^2 - v^2\right) - v^2}{c^2 - v^2} \right\} \qquad \text{(vii)}$$

$$\rightarrow P_d = 4L \left\{ \frac{c\sqrt{c^2 - v^2} - c^2}{c^2 - v^2} \right\} \qquad \text{(viii)}$$

And the fringe shifts will equal to $S_f = P_d / \lambda$ $\qquad$ (ix)

$$\rightarrow S_f = P_d / \lambda = \frac{4L}{\lambda} \left\{ \frac{c\sqrt{c^2 - v^2} - c^2}{c^2 - v^2} \right\} \qquad \text{(x)}$$

For $L = 32m$, $v = 30km/s$, $c = 300,000km/s$, $\lambda = 640nm$.

$$\rightarrow S_f \approx 1.0 \ \underline{\text{fringe shift}}$$

If the arm length equals to 32 meters and the wavelength of the He-Ne laser equals to 640nm, while as the interferometer was rotated for 90° CCW, there will be about one fringe shift of the interference pattern.

However, the fringe shift of the experiment result is less than 1% of what expected or almost zero. This result was attributed to the experimental inaccuracy error. More scientists have engaged into similar experiments with increasing sophistication since then. However, their results were all the same.

Name	Year	Arm length	Fringe shift expected	Fringe shift measured
Michelson	1881	1.2	0.04	0.02
Michelson and Morley	1887	11	0.4	< 0.01
Morley and Miller	1902–1904	32.2	1.13	0.015
Miller	1921	32.0	1.12	0.08
Miller	1923–1924	32.0	1.12	0.03
Miller (Sunlight)	1924	32.0	1.12	0.014
Tomascheck (Starlight)	1924	8.6	0.3	0.02
Miller	1925–1926	32.0	1.12	0.088
Kennedy (Mt Wilson)	1926	2.0	0.07	0.002
Illingworth	1927	2.0	0.07	0.0002
Piccard and Stahel (Rigi)	1927	2.8	0.13	0.006
Michelson et al.	1929	25.9	0.9	0.01
Joos	1930	21.0	0.75	0.002

(Table3.3-1: List of experiments in history)

3.4 Aberration of Light on Interferometer

3.4.1 Aberration of Light Beams

The work and concept of the Michelson–Morley experiment seems to be so reasonable and unquestionable. However, there was something really important but have been left out so far which created a huge aberration in our history. We could see how far this aberration was if we re-analyze this experiment all over again from another point of view.

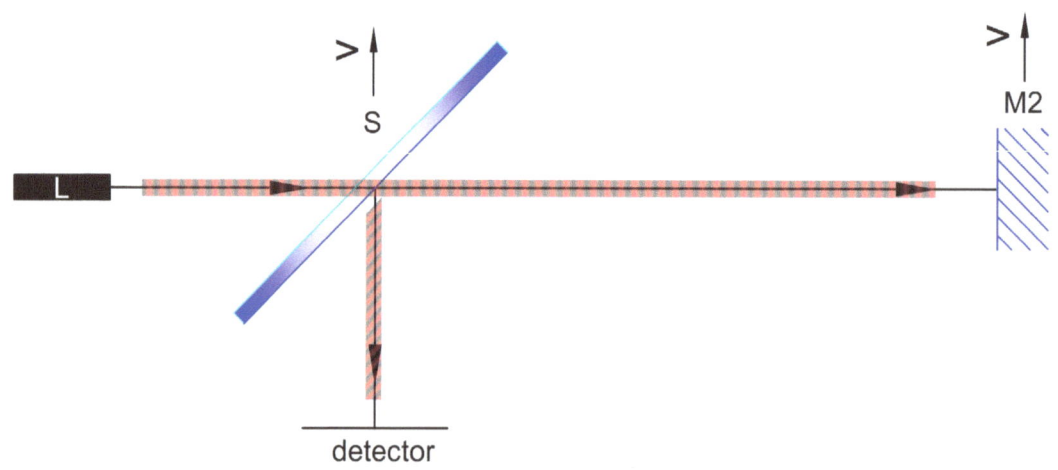

(Fig3.4-1: **Beam-B** on a moving interferometer)

Below we will re-analyze the path of **beam-B** in more detail. Since $m1$ is moving, $m2$, S and the whole interferometer are also moving at the same time. When **beam-B** left from S to $m2$, $m2$ is moving upwards. When **beam-B** was reflected from $m2$ back to S, S is also moving upwards. If we take the whole interferometer as stationary, the path of **beam-B** will become as illustrated below (Fig3.4-2) for the velocity of light wave should have nothing to do with its source.

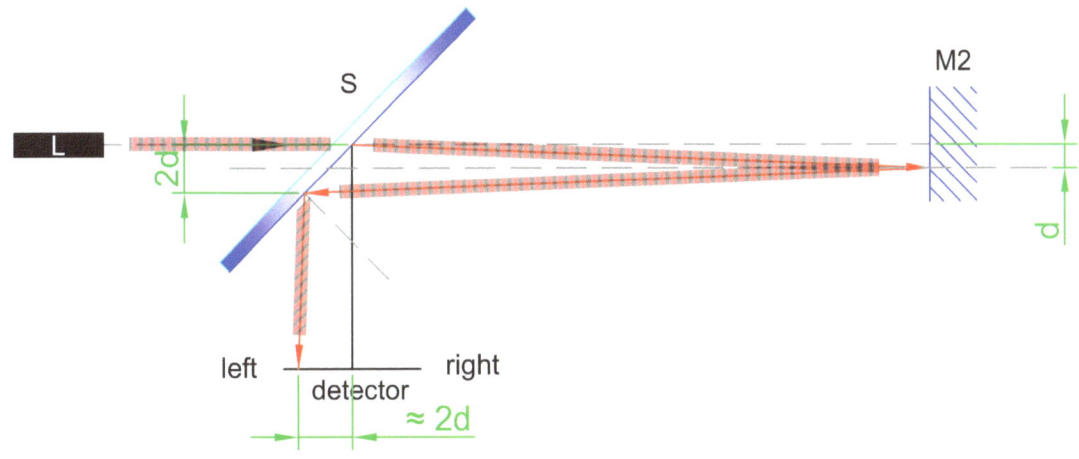

(Fig3.4-2: **Beam-B** with respect to stationary interferometer)

Where (Fig3.4-2) can be simplified as below drawing (Fig3.4-3):

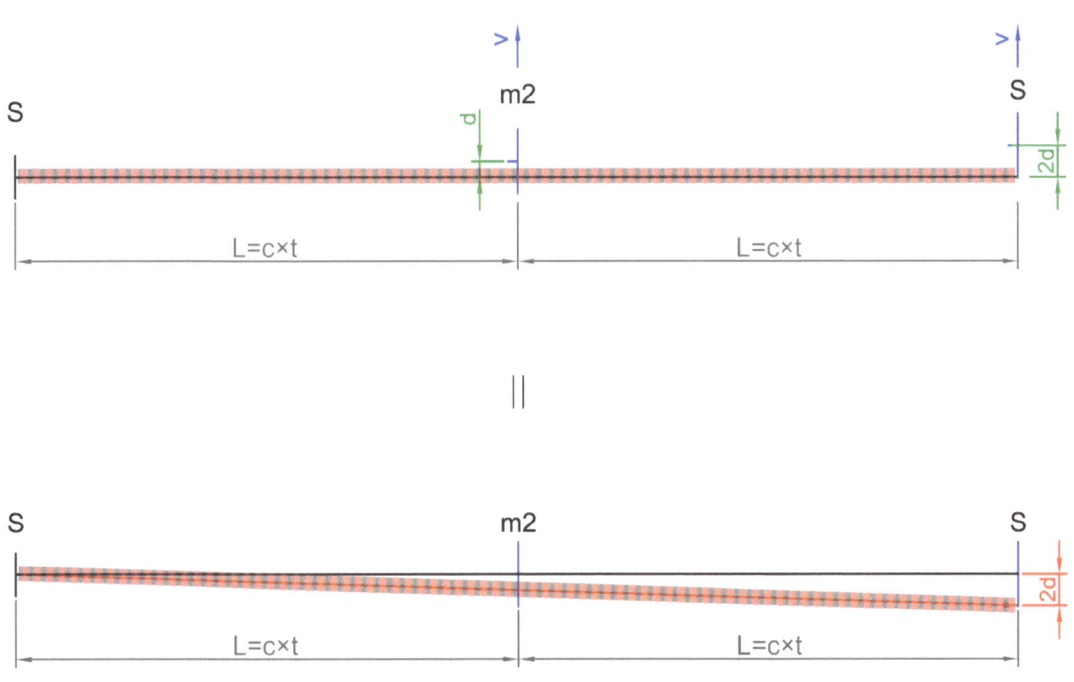

(Fig3.4-3: Simplified path of **beam-B**)

This is actually the effect of light aberration. If light beam moved for a distance of $2L$ horizontally, it will also move for a distance of $2d$ vertically with respect to the interferometer.

$$\therefore L = c \times t, \quad d = v \times t$$

$$\Rightarrow t = \frac{L}{c} = \frac{d}{v}$$

$$\rightarrow d = \frac{v}{c} \times L$$

For $\frac{v}{c} \approx 10^{-4}, \quad L = 32m$

$$\rightarrow 2d = 2 \times \frac{v}{c} \times L = 2 \times 10^{-4} \times 32$$

$$= 6.4 \times 10^{-3} m = 6.4mm$$

Such value of the displacement of the light pattern is normally observable by the naked eyes.

While as the interferometer was rotated 360° around, the relative shift of the projection between **beam-A** and **beam-B** will equal to $2d$ between each phase in order shown as below drawing (Fig3.4-4).

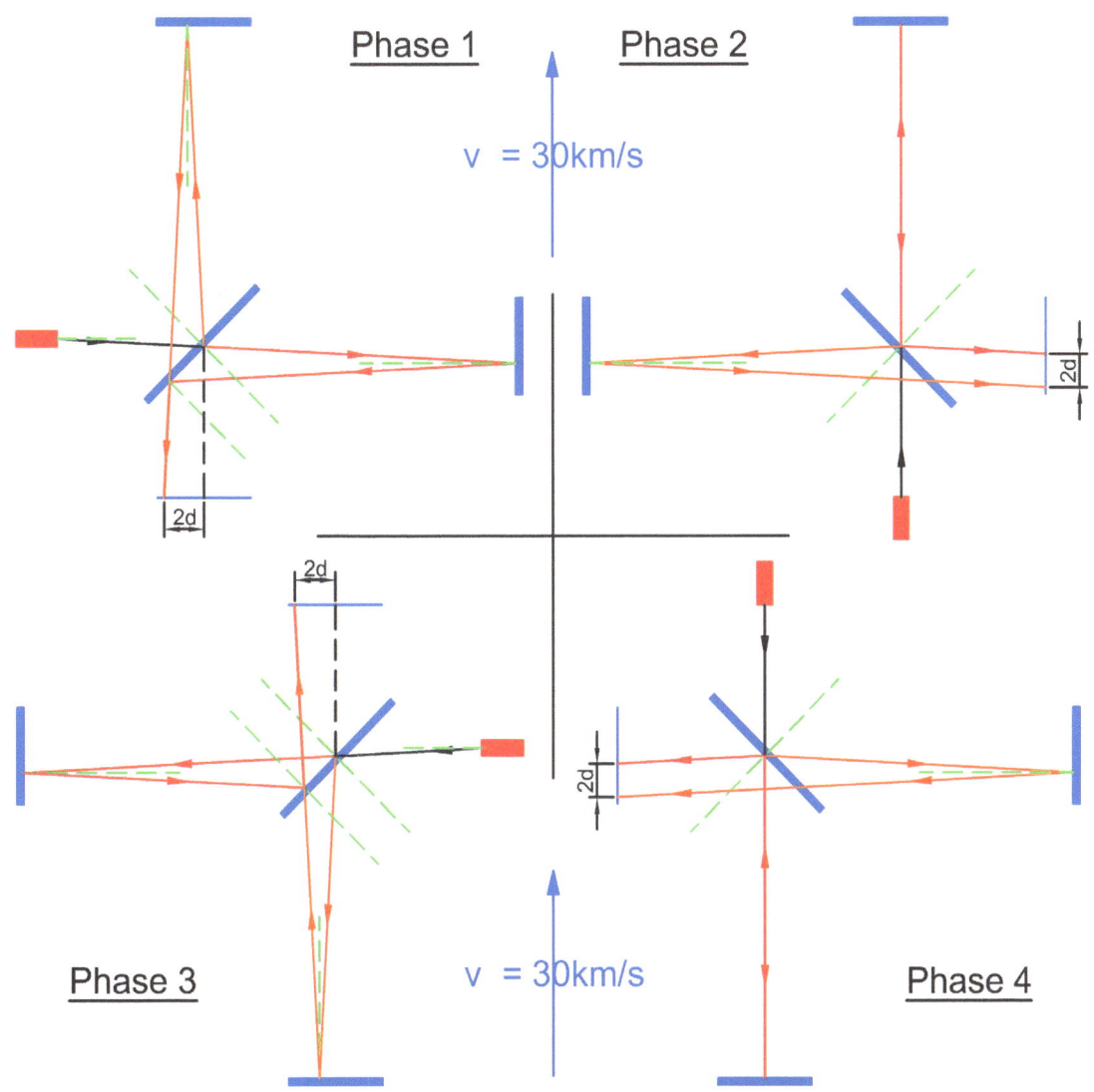

(Fig3.4-4: Relative shift of the projections between **beam-A** and **beam-B**)

The key is that the projection of the light beams on the detector were transverse placed. It is similar to projecting two similar pictures on the same screen. If the projections were shifted transversely, it is equivalent to shifting of the interference fringes. However, this did not happen actually. The projection of light beams did not shift, which means there is no aberration of the light beams from the interferometer.

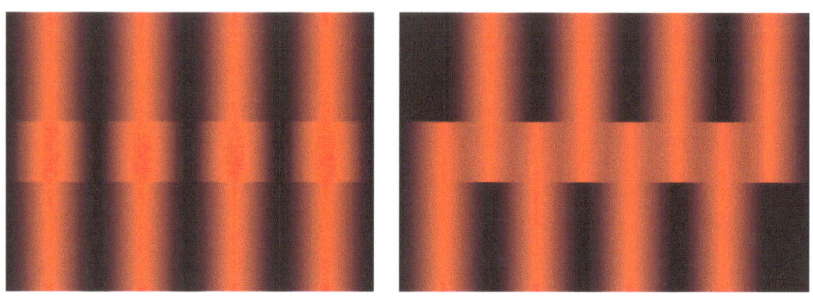

(Fig3.4-5: Relative shift between two similar pictures)

And hence the fringe shift will equal to $2d/\lambda$.

$$\rightarrow 2d/\lambda = \frac{6.4mm}{640nm} = \frac{6.4\times10^{-3}}{640\times10^{-9}} = 10^4 \quad \underline{\text{fringe shifts}}$$

Although we can use a magnifier in front of the detector to compensate the shifting of the light pattern, however, since the result is quite large, it should be very easy to detect such shift by naked eyes and the light patterns could even shift beyond the scope of the detector. However, there is no such experiment phenomenon in fact.

3.4.2 Experiment of Aberration of Light

Since the shift of the projection has an amount of mini-meters, it's easy to perform an experiment to test it. We can use simple equipments to construct this experiment as following:

1) A rigid bar of 6 meters long and a mark-place on it.
2) A commercial LED pointer.
3) A pen and a note paper.

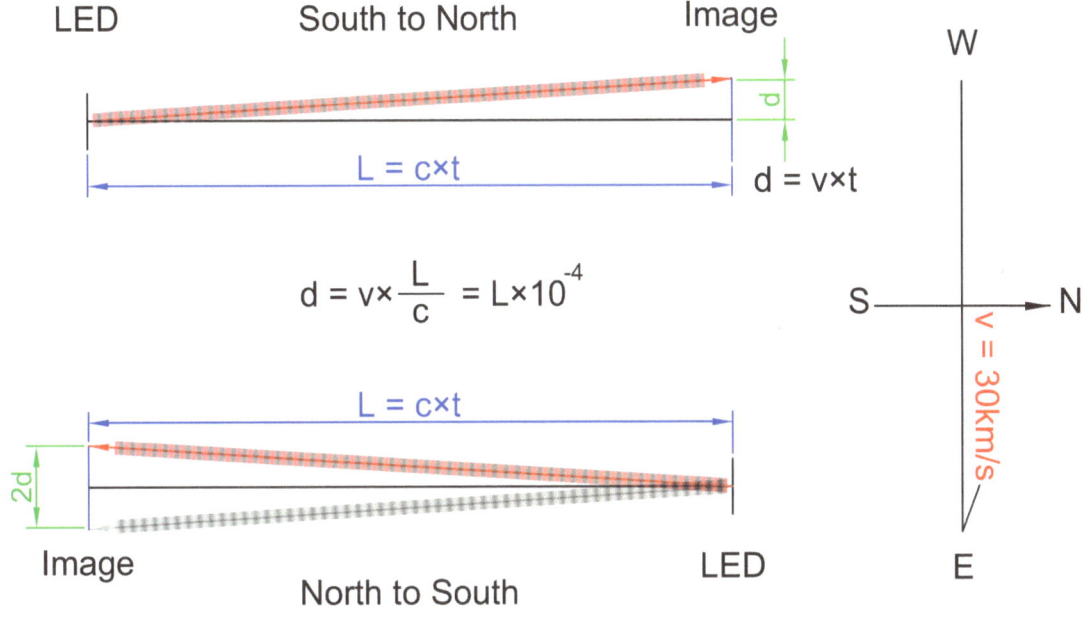

(Fig3.4-6: Experiment of light aberration)

At 12AM or 12PM, place the bar north-to-south. Fix the note paper on the mark-place at one end (north end) of the Aluminum bar. Then fix the LED pointer on the other end (south end) and let it project right on the note paper. The LED light beam will emit from south to north. Record or mark the projection position.

Rotate the bar for 180° horizontally. Let the LED light emit from north to south. Check and compare the projection position of LED light beam with previously marked.

For L = 6M, the expected shift of projection should be more than 1mm at noon or midnight. However, we did not find such phenomenon.

3.4.3 Examples to Question on Aberration of Light

A) The Global Positioning System (GPS)

The Global Positioning System (GPS) is the only fully functional Global Navigation Satellite System (GNSS). The GPS uses a constellation of between 24 and 32 Medium Earth Orbit satellites that transmit precise electromagnetic wave signals, which enable the GPS receivers to determine their current location, the time, and their velocity (including direction).

The GPS satellite orbits Earth at an altitude of approximately 20,200 kilo-meters. The GPS provides an accuracy of about 10m or better in positioning. While as the GPS satellite transmitting electromagnetic wave signal to ground receivers, we have to consider several effects and errors such as the *atmospheric effects, multipath effects, ephemeris and clock errors, geometric dilution of precision computation (GDOP), Sagnac effect and some other sources of interference* during transmitting to correct the positioning. However, the effect of aberration of light is omitted and seems to be unnecessary.

Since the orbital motion of Earth around the Sun has a speed of about 30km/s, the time for the signal transmitted from the GPS satellite to the receiver will be about 0.1sec. Thus the horizontal distance error could be kilo-meters for the same receiver at different time (i.e. v = 0 at 6AM and v = 30km/s at 12AM), and the present high-precision GPS would be entirely impossible.

The aberration error causes on the GPS signals can be shown as the following demonstration:

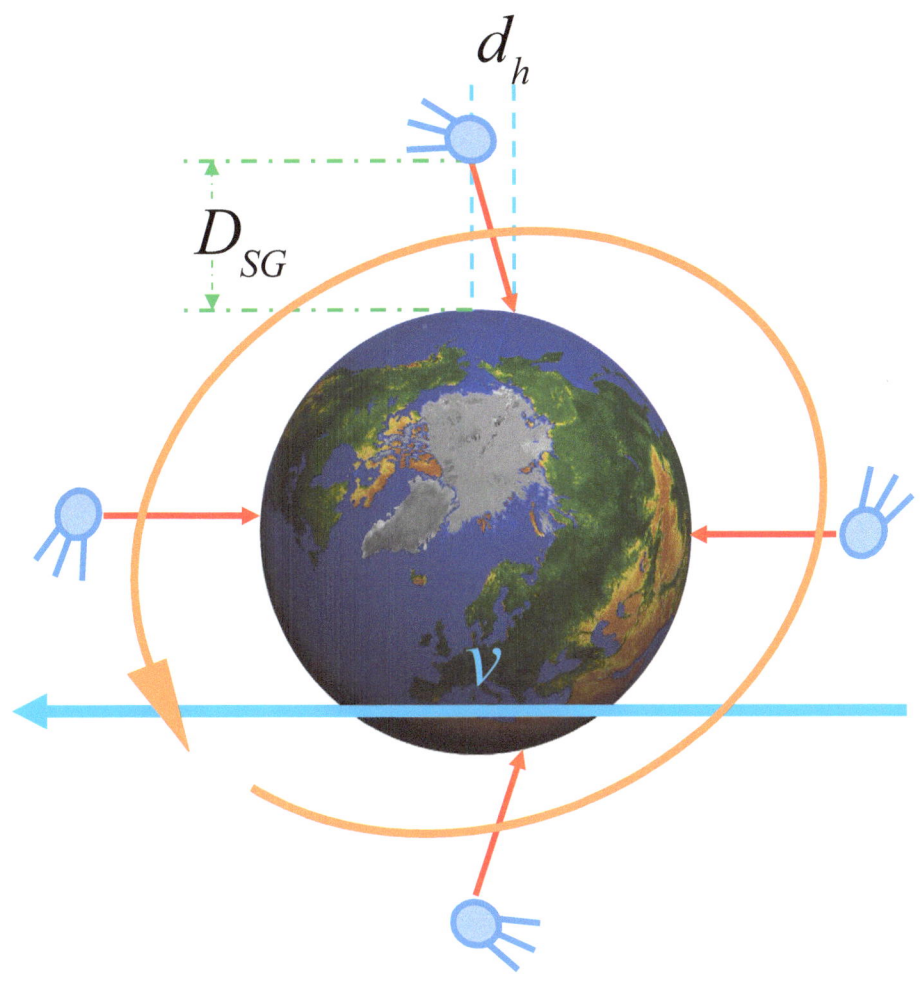

(Fig3.4-7: GPS satellites revolving around Earth)

$$\rightarrow d_h = D_{SG} \times \frac{v}{c}$$

$$\because D_{SG} = 20,000 km = 2 \times 10^4 km$$

$$\rightarrow d_h = 2 \times 10^4 \times 10^{-4} = 2km$$

B) The CDs and DVDs

The CDs and DVDs are the more practical examples to question on the light aberration. The width of the data tracks on the CD and DVD disc is less than 0.5μm. While the laser beam travel back and forth between the laser head and the disc surface, the aberration of light can be calculated as below:

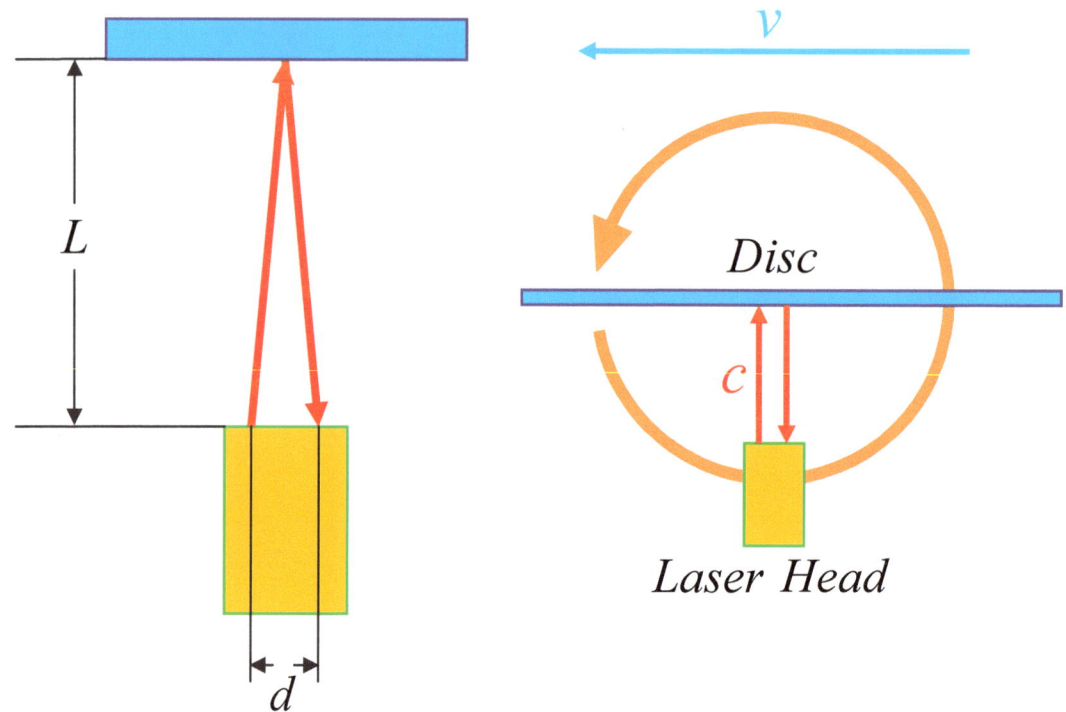

(Fig3.4-8: CD player and the disk)

$$d = 2L \times \frac{v}{c} \qquad , \; 2L = 1cm = 10^{-2}m$$

$$\rightarrow d = 10^{-2} \times 10^{-4} = 10^{-6}m = 1\mu m$$

Thus, our CDs and DVDs will not work all day long since the direction of the velocity of the CD and DVD players with respect to the laser beam can vary freely in all direction.

3.5 The Velocity of Light on Earth

The hypothesis of stellar aberration by James Bradley seems to be the simplest and the most practicable explanation for the astronomical phenomenon between stars and the observers. However, if superposition of velocity is the cause of stellar aberration, change of light speed would be detectable on Earth. If change of transverse component of velocity is the cause of light aberration, it should happen to all the lights at any place. However, change of light speed has not been detected on Earth, and aberration of light does not happen to those lights whose source locates on Earth.

3.5.1 Superposition of Velocity
The combination of the velocity according to the rule of the superposition of the velocity should comply with the following *Cosine Formula*:

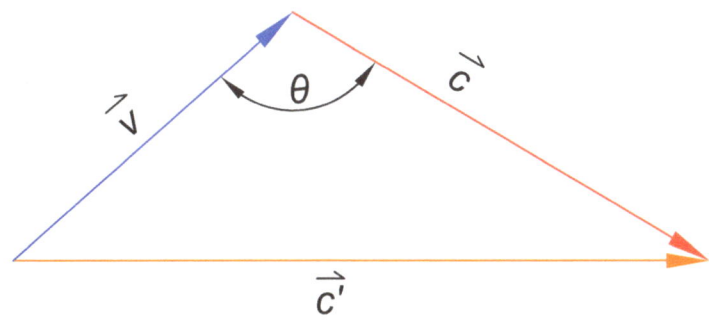

$$c' = \sqrt{c^2 + v^2 - 2c \cdot v \cdot \cos\theta}$$

(Fig3.5-1: Superposition of velocity with Cosine Formula)

3.5.2 The Velocity of Earth in the Universe

The rotation speed of Earth could be omitted, because it's not fast enough comparatively to others. The revolution velocity of Earth around the sun is about 30km/s. And the revolution velocity of the sun around the Galaxy is about 250km/s. Which means the relative velocity of the Earth with respective to the Galaxy could be about 280km/s. Even if we do not consider the revolution velocity of the Galaxy around the **LG** (**Local Group**) or the **LG** around the **LS** (**Local Supercluster**), the relative velocity of the Earth with respect to the Galaxy will make the light-speeds of the stars we detect to have maximum ±280km/s difference at different time and different position.

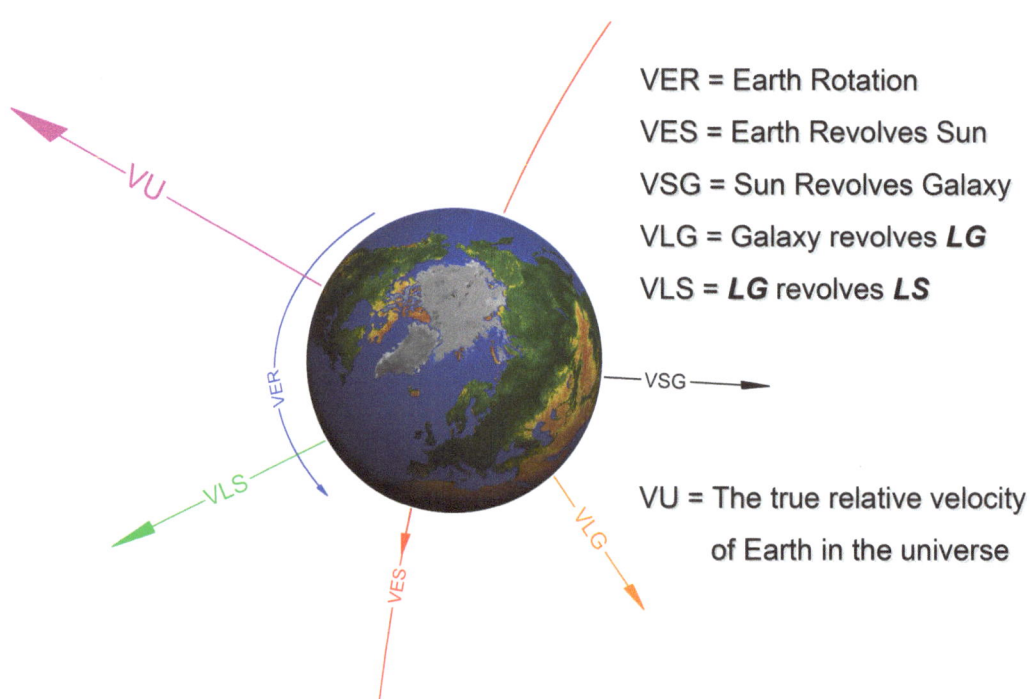

VER = Earth Rotation
VES = Earth Revolves Sun
VSG = Sun Revolves Galaxy
VLG = Galaxy revolves **LG**
VLS = **LG** revolves **LS**

VU = The true relative velocity of Earth in the universe

(Fig3.5-2: Relative velocity of Earth with respect to each system)

Above has not yet taking consideration of the revolution velocity of the Galaxy around the **LG**, the revolution velocity of the **LG** around the **LS**, and the revolution velocity of the **LS** around the universe. In other words, the actual relative velocity of Earth with respect to the universe could possibly be very high (may even higher than light speed).

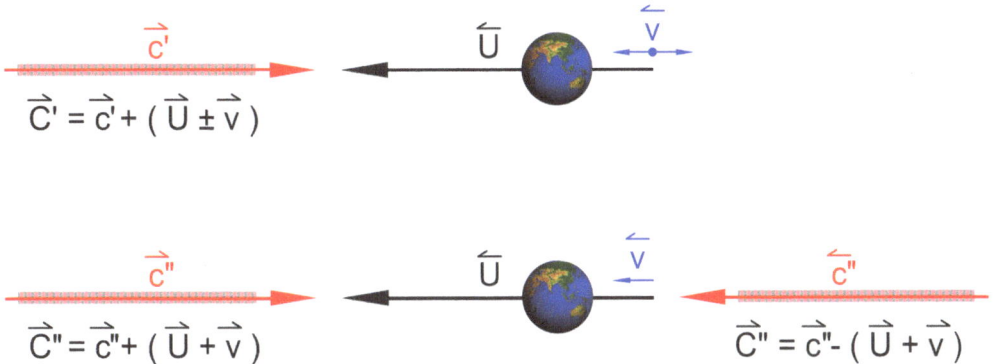

$$\vec{C}' = \vec{c}' + (\vec{U} \pm \vec{v})$$

$$\vec{C}'' = \vec{c}'' + (\vec{U} + \vec{v})$$

$$\vec{C}'' = \vec{c}'' - (\vec{U} + \vec{v})$$

(Fig3.5-3: Relative velocity of Earth with respect to light)

As illustrated above, assume the true velocity of the Solar System with respect to the universe as $\overline{U}$, and the revolution velocity of Earth around the sun as $\vec{v}$. For starlight, change of velocity could be only $\vec{v}$, because the velocity and direction of starlight does not change, and so doesn't $\overline{U}$. This situation could be represented as top of (Fig3.5-3).

However, if the direction of starlight was turned for 180°, the velocity of Earth in the universe will become $\overline{U} + \vec{v}$ with respect to starlight. Since $\vec{v}$ is about 30km/s, and it is possible that $\overline{U}$ could be thousands or hundreds of thousand times of $\vec{v}$, in this case, what if the direction of light and $\overline{U}$ is perpendicular to each other? This is just the same case in the Michelson–Morley experiment and the experiment of light aberration.

The speed change in the Michelson–Morley experiment could never be the revolution velocity of Earth around the sun. It must be much greater than 30km/s ($\overline{U} + \vec{v}$). However, this doesn't make any difference anyway. The results for the Michelson–Morley experiment and all similar experiments are all the same. The relative speed of light with respect to everything on Earth is always c. In this case, for the experimental light beams c'' on Earth can be displayed as the following equations:

$$\vec{c''} + \left(\vec{U} + \vec{v} \right) = \vec{c}$$

$$\rightarrow \vec{c''} = \vec{c} - \left(\vec{U} + \vec{v} \right)$$

This means except for the velocity of light, the light beam in the experiment should have another component of velocity with respect to the universe which is the same with its source and all other parts in the experiment on Earth. This complies with our postulation of **Case A** in *Chapter 2 Part 2.2.3*. There is no relative velocity between Earth and the medium of light on Earth.

3.6 Gravitational Space Vortex

3.6.1 Gravitation and Planetary Motion

As we all know that it is the Gravitation which upholds the sun and the planets of the Solar System together and makes those planets to revolve around the sun in circular motions. And the gravitational force is an attractive force, its direction is always pointing towards the center of the mass. However, to cause a circular motion, a force on the circle which always points towards the tangent of the object on the circle is required. While a circular motion was formed, the circulation circle will become invariant only when the generated centrifugal force and the centripetal force (gravitation) counterbalance with each other.

In the mean time, all the planets, asteroids, meteorites and other celestial objects in the Solar System in orbit around the Sun lie near the plane of ecliptic. And all of the planets and most other objects also orbit with the Sun's rotation (counter-clockwise, as viewed from above the Sun's north pole). Such situations cannot be simply explained only by Gravitation. There must be a certain force on the ecliptic which causes all the celestial objects to revolve around the sun.

We can describe the gravitational acceleration caused by the sun in the planetary motion as following:

1) The centripetal acceleration (gravitational acceleration) caused by the sun for all the objects in the Solar System can be described as:

$$a = \frac{GM}{r^2} \qquad \text{(i)}$$

2) When a celestial object orbits around the sun in a circular motion (or its orbit is almost a circle), its centripetal acceleration can be simplified as:

$$a = \frac{v^2}{r} \qquad \text{(ii)}$$

Combine (i) and (ii)

$$a = \frac{GM}{r^2} = \frac{v^2}{r} \qquad \text{(iii)}$$

$$\Rightarrow v = \sqrt{\frac{GM}{r}} \qquad \text{(iv)}$$

Where G is the gravitational constant; M is mass of the sun; r is the distance between the object and the sun; v is the speed of the object traveling in the circle and also known as the revolution velocity around the sun.

3.6.2 Gravitational Vortex
The result ($v = \sqrt{GM/r}$) has nothing to do with the mass of the object itself.

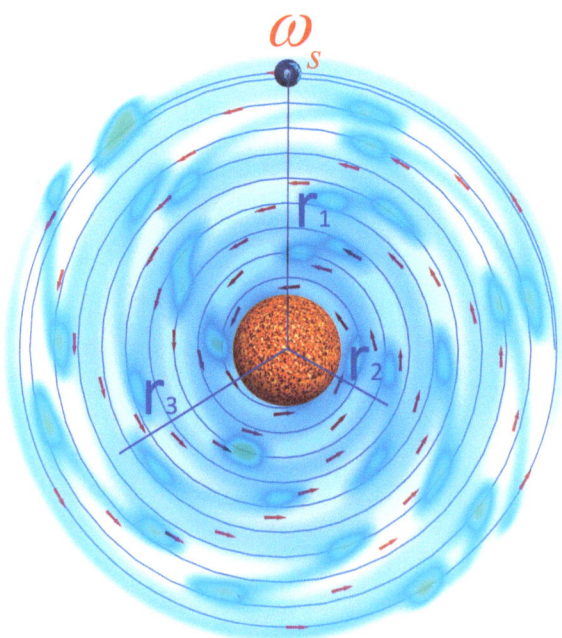

(Fig3.6-1: Gravitational vortex of the Solar System)

This can be taken as everything including the space (no matter it has mass or not) is circling around the sun in a speed relative to its position with respect to the gravitation center of the sun. In this case, we can imagine that the space is flowing around the sun as a free vortex since the angular speed of the space at the outer ring of the vortex is always slower than which at the inner side.

$$v = \omega_s \cdot r \quad \Rightarrow \quad \omega_s = \frac{v}{r} \tag{v}$$

$$\rightarrow \omega_s = \sqrt{\frac{GM}{r}} \Bigg/ r = \sqrt{\frac{GM}{r^3}} \tag{vi}$$

$$\Rightarrow \omega_s \propto \frac{1}{\sqrt{r^3}} \tag{vii}$$

As Earth drifting in the space for an extreme very long time, Earth will move along with the flow of space in the same velocity. In other words, the relative velocity of Earth with respect to the flow of space is zero.

For:
1) The relative velocity of Earth with respect to the space flow is zero.
2) The relative velocity of light with respect to its medium is constant c.

In this case, the light speed with respect to Earth is constant c as well. This can be taken as Earth is stationary with respect to the space in its vicinity. Thus the revolution speed of Earth around the sun can be omitted or taken as 0 for the light beams on Earth. This complies with our postulation of **Case A** in *Chapter 2 Part 2.2.3.* and all the experiment results and the phenomena of light on Earth.

3.6.3 Starlight Aberration and Space Flow

For the starlight comes from the outer space into the Solar System and approaching towards the sun or Earth, the every step of the photon which it makes for its propagation with respect to the speed change of the space flow can be simplified as the following pictures.

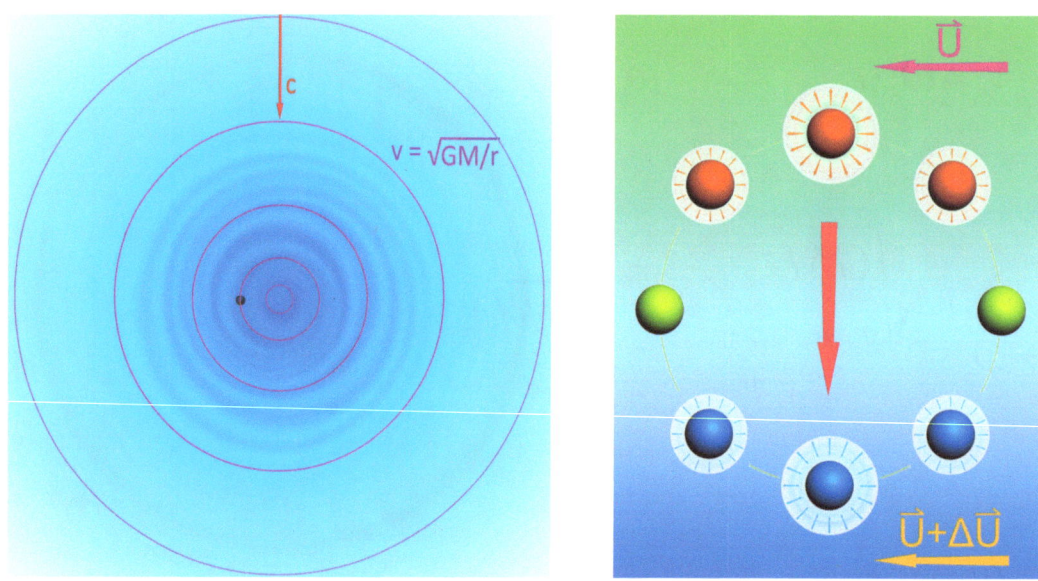

(Fig3.6-2&3: Starlight and the speed difference of the space flow)

If we magnify to see a photon as stationary, the relative velocity of the space flow with respect to the static photon will be a little difference ahead and astern of the photon. The difference of the relative velocity $\Delta \vec{U}$ of the space is very small and almost zero since the photon is extremely small.

When a photon moves in the space, the drag effect of the space to the photon is considered as it is a particle drifting the space, and hence can be considered as the acceleration to the photon. However, as far as the propagation speed of the photon is concerned, the photon is considered as it is a kind of waveform propagates in the space. And hence the relative velocity of the photon with respect to its vicinity space will be limited to the light speed in that medium. Which shall be the constant c in the vacuum.

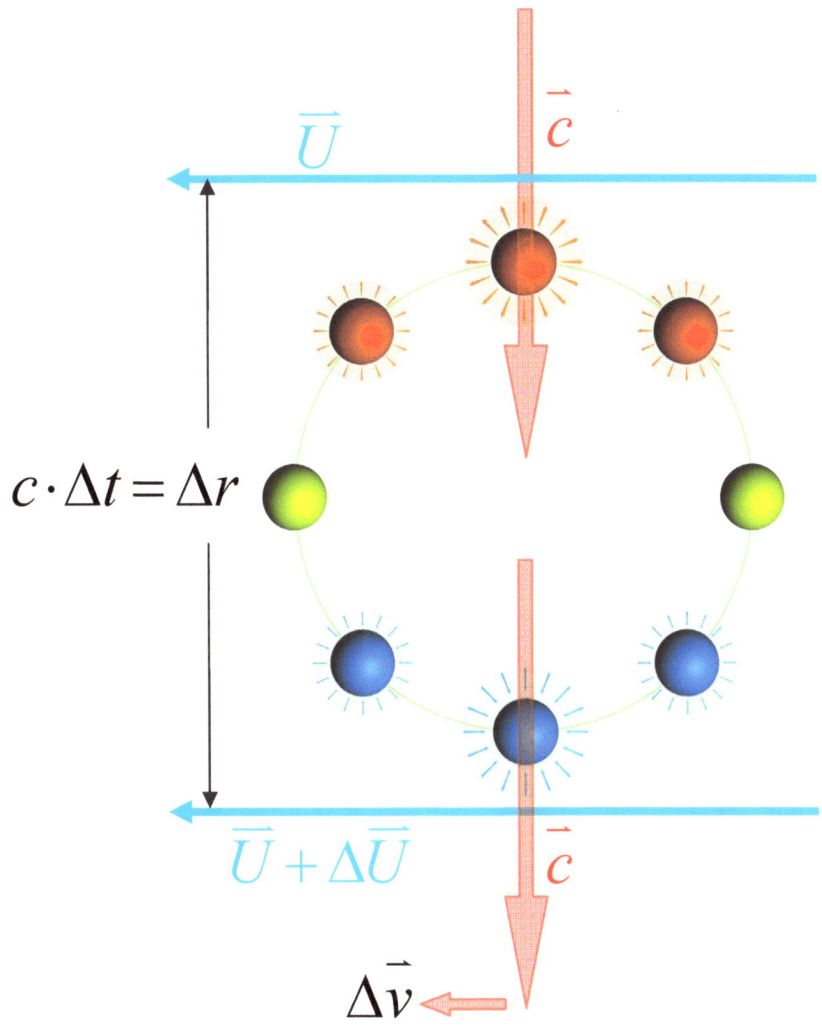

$$c \cdot \Delta t = \Delta r$$

(Fig3.6-4: Speed difference of the space flow and the photon)

The space flow may cause the drag effect to the photon. However, because the drag coefficient is very small and the difference of space flow is also very small, this will result in a zero-drag of the space to the photon. In this case, the flow of the space will not take any effect on the propagation of the photon. The acceleration caused by the space flow to the photon can be described as:

$$a = C_x \cdot \Delta U \qquad \text{(i)}$$

Note that there will not be any drag effect if the speed of space flow is invariant. The acceleration only happens when the photon passing through different space flow. The speed increase Δv caused by the acceleration can be described as the following equation:

$$\Delta v = a \cdot \Delta t = C_x \cdot \Delta U \cdot \Delta t \qquad \text{(ii)}$$

Let

$$U = \sqrt{\frac{GM}{R+(n+1)\Delta r}} \quad \text{and} \quad U + \Delta U = \sqrt{\frac{GM}{R+n\Delta r}} \qquad \text{(iii)}$$

$$n \in Z > 0$$

$$\because \Delta U = (U + \Delta U) - U \qquad \text{(iv)}$$

$$\Rightarrow \Delta U = \sqrt{\frac{GM}{R+n\Delta r}} - \sqrt{\frac{GM}{R+(n+1)\Delta r}} \qquad \text{(v)}$$

$$\Rightarrow \Delta v = C_x \left(\sqrt{\frac{GM}{R+n\Delta r}} - \sqrt{\frac{GM}{R+(n+1)\Delta r}} \right) \Delta t \qquad \text{(vi)}$$

Meanwhile, the propagation speed of the photon in the space is caused by the time-space vibration of the halftons, the propagation speed of the photon with respect to its medium will be limited to the speed of light wave in that medium. The speed of the photon will be as the constant c. Thus, $\Delta r = c \times \Delta t$.

$$\therefore \Delta t = \Delta r \big/ c$$

$$\rightarrow \Delta v = C_x \left(\sqrt{\frac{GM}{R+n\Delta r}} - \sqrt{\frac{GM}{R+(n+1)\Delta r}} \right) \frac{\Delta r}{c} \qquad \text{(vii)}$$

$$\rightarrow \Delta v = \frac{C_x \sqrt{GM}\,\Delta r}{c} \left(\sqrt{\frac{1}{R+n\Delta r}} - \sqrt{\frac{1}{R+(n+1)\Delta r}} \right) \qquad \text{(viii)}$$

For a photon moves from an infinite distance away from the Solar System and reached Earth or reversely, the total speed increase v_t will be the summation of Δv which can be described as the following:

$$v_t = \sum \Delta v \qquad \text{(ix)}$$

$$\rightarrow v_t = \sum_{n=0}^{n=\infty} \frac{C_x \sqrt{GM}\,\Delta r}{c} \left(\sqrt{\frac{1}{R+n\Delta r}} - \sqrt{\frac{1}{R+(n+1)\Delta r}} \right) \qquad \text{(x)}$$

Extend above equation as following:

$$\rightarrow v_t = \frac{C_x \sqrt{GM}\,\Delta r}{c}\left(\sqrt{\frac{1}{R+0\Delta r}} - \sqrt{\frac{1}{R+(0+1)\Delta r}}\right.$$
$$+\sqrt{\frac{1}{R+1\Delta r}} - \sqrt{\frac{1}{R+(1+1)\Delta r}}$$
$$+\sqrt{\frac{1}{R+2\Delta r}} - \sqrt{\frac{1}{R+(2+1)\Delta r}}$$
$$\cdot$$
$$\cdot$$
$$\cdot$$
$$\left.\sqrt{\frac{1}{R+\infty\Delta r}} - \sqrt{\frac{1}{R+(\infty+1)\Delta r}}\right)$$

(xi)

$$\rightarrow v_t = \frac{C_x \sqrt{GM}\,\Delta r}{c}\left(\sqrt{\frac{1}{R+0\Delta r}} - \sqrt{\frac{1}{R+(\infty+1)\Delta r}}\right) \quad \text{(xii)}$$

$$\Downarrow$$
$$0$$

$$\rightarrow v_t = \frac{C_x \sqrt{GM}\,\Delta r}{c}\left(\sqrt{\frac{1}{R}} - 0\right) \quad \text{(xiii)}$$

$$\rightarrow v_t = \frac{C_x \Delta r}{c} \sqrt{\frac{GM}{R}} \qquad \text{(xiv)}$$

When both of the halftons revolving around their mass center, the average distance between them in their moving direction equals to the radius of the photon, which equals to the wave length of the light beam. Thus the minimum value of Δr can not be less than λ.

$$\Rightarrow v_t = \lim_{\Delta r \to \lambda} v_t \qquad \text{(xv)}$$

$$\rightarrow v_t = \lim_{\Delta r \to \lambda} \frac{C_x \Delta r}{c} \sqrt{\frac{GM}{R}} \qquad \text{(xvi)}$$

$$\rightarrow v_t = \frac{C_x \lambda}{c} \sqrt{\frac{GM}{R}} \qquad \text{(xvii)}$$

Where $\sqrt{\frac{GM}{R}} = V_E$ is the revolution speed of Earth around the sun.

$$\rightarrow v_t = \frac{C_x \lambda}{c} \cdot V_E = C_x \cdot \lambda \cdot \frac{V_E}{c} \qquad \text{(xviii)}$$

Where $C_x \leq 1/s$, $\lambda \leq 1\mu m$, $V_E \Big/ c \approx 10^{-4}$

$$\to v_t \leq 1 \times 10^{-6} \times 10^{-4} = 10^{-10} m/s \approx 0 \qquad \text{(xix)}$$

Thus, when a photon moves in the space with small variant velocity of the space flow, the moving direction of the photon in the universe will not be affected by the flow of the space if the difference of the space flow is extremely small. And the propagation speed of the photon with respect to the space will always be invariant as the light speed in that medium. As a result, this will change the relative velocity of light with respect to the space flow but remain the relative speed of light wave with respect to the space. That is to say, the starlight will remain its propagation speed as constant c but change its propagation direction with respect to the flow of the space.

For example, before the starlight enters the Solar System, the velocity of starlight with respect to the Solar System is $\vec{c}$, and it's perpendicular to the velocity of the Solar System in the universe.

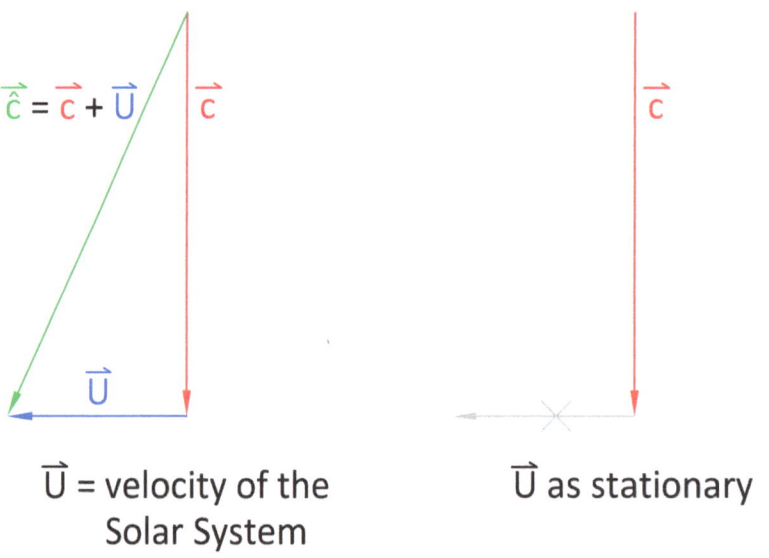

$\vec{U}$ = velocity of the Solar System

$\vec{U}$ as stationary

(Fig3.6-4: Starlight before entering the Solar System)

In this case, if we take the Solar System as stationary, the direction of the starlight will still be perpendicular with respect to the Solar System. The relative speed of light with respect to the space is constant c.

After the starlight enters the Solar System, the velocity of the space flow will start to change while as the direction of the starlight will remain the same. In this case, if we take the local flow of the space as stationary, the starlight will have an extra component of velocity which is opposite to the flow of space.

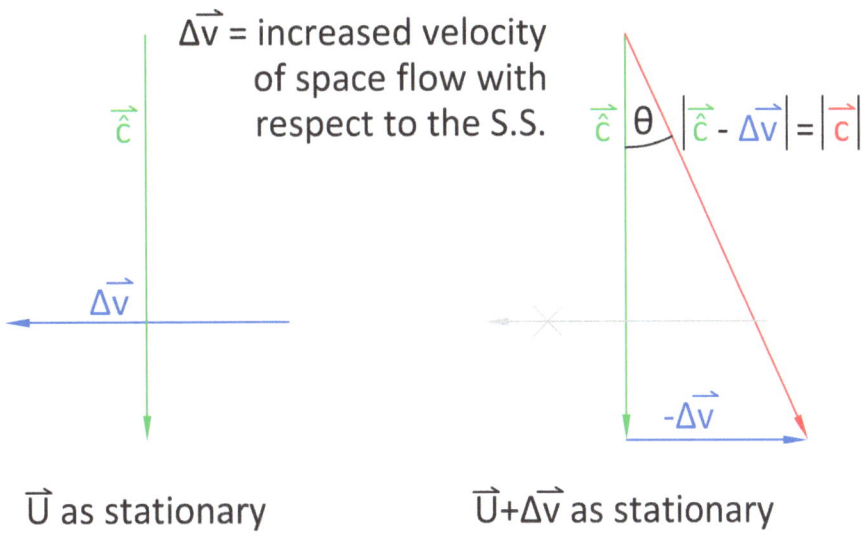

$\vec{\Delta v}$ = increased velocity of space flow with respect to the S.S.

$\vec{U}$ as stationary

$\vec{U}+\vec{\Delta v}$ as stationary

(Fig3.6-5: Starlight after entering the Solar System)

Thus, the starlight will become tilted. However, the relative speed of the starlight with respect to the flow of space will remain as light speed in that medium. And the inclination angle can be described as below:

$$\theta = \arcsin\left(\frac{\Delta v}{c}\right).$$

This is actually what we named the Stellar Aberration in *Part 3.2*.

3.6.4 Spiral Ether Model

Before we redefine the Ether model, we have to clarify something. First of all, no, there is no such thing named Ether in the space. Ether is redundant for our theory. The second, space is just space, however, it does possess the property of dimension. Ether model is only a concept to describe the space as a medium of light and all other particles. We keep adopted the Ether concept to honor the greatness of the ancient geniuses.

According to the phenomena of the planetary motion and the astronomical observation, the Solar System, the Galaxy, and the whole universe are spirals. In this case, we should modify the "*Stationary Ether Model*" into the "*Spiral Ether Model*" instead.

It was said that if Earth moves in the stationary Ether, there will be some kind of the friction between Earth and Ether, and will finally slow down the speed of Earth revolving around the sun.

In the case of an object in a water vortex, the object with respect to the water at its vicinity can be considered as static. There is no relative velocity between the object and the water around the object. It is the water which actually moves that drags the object to drift around the vortex center.

The gravitational vortex is quite similar to the water vortex. We can take the gravitational field of the Solar System as a giant vortex, and the planets are small vortices in the giant vortex, and our moon and those satellites of other planets are smaller vortices. What flows within the vortices is the space.

For the Solar System, the gravitational vortex of the sun will drag the space to revolve around and flow into the vortex center of the sun. The planets in the Solar System will be dragged by the space flow cause by the sun, and makes the planets to revolve around the sun and to be attracted by the sun. When the centrifugal force generated by the revolution of the vortex and the centripetal force generated by the space flowing into the vortex center counterbalance with each other, the planetary motions will last.

CHAPTER 4

DISCUSSIONS

4.1 Is Light Speed Constant or Not

Is light speed constant or not? This is a question debated by the scientists for over a century. Actually, it's a simple question and the answer is right there. But somehow people are just blind to see.

We know that the light speed with respect to any stationary objects on Earth is constant in all direction. However, in some equipment which is not stationary on Earth, the light speed with respect to the equipment is not constant. Hereby we are going to discuss some phenomena and effects which disprove the proposition of the constancy of the speed of light.

4.1.1 The Space Dragged by Ground

Since there is no relative velocity between Earth and the space vortex of the sun, hence the revolution speed of Earth around the sun can not affect the light beams on Earth. Meanwhile, since all of the objects and even the air are moving synchronously with the rotation of Earth, hence everything on the ground surface including the space inside the atmosphere can be taken as a closed system. This means the space inside the atmosphere is dragged synchronously by Earth rotation. The space inside the atmosphere will not have any relative velocity with respect to the atmosphere and ground. And hence the relative velocity of light with respect to everything inside the atmosphere which is stationary with respect to the ground surface will be the same in all direction.

The space inside the atmosphere is dragged by Earth rotation synchronously. But the space flow outside the atmosphere will follow the equation ($v = \sqrt{GM/r}$) derived in *Chapter 3 Part 3.6*. In this case, some experiments like the Michelson–Morley experiment which has been performed on the ground surface, stationary with respect to ground, can never detect the relative velocity of the medium of light since there is no relative velocity of the space with respect to the ground surface. And the result will be the same even outside the atmosphere. This complies with all the experiment results on Earth.

4.1.2 Sagnac Effect

The Sagnac effect (also called as Sagnac interference), named after French physicist Georges Sagnac, is a phenomenon encountered in interferometry that is elicited by rotation. The time difference Δt between the light beams travel on a Sagnac interferometer CW and CCW is proportional to the loop area A and the angular speed ω of the interferometer.

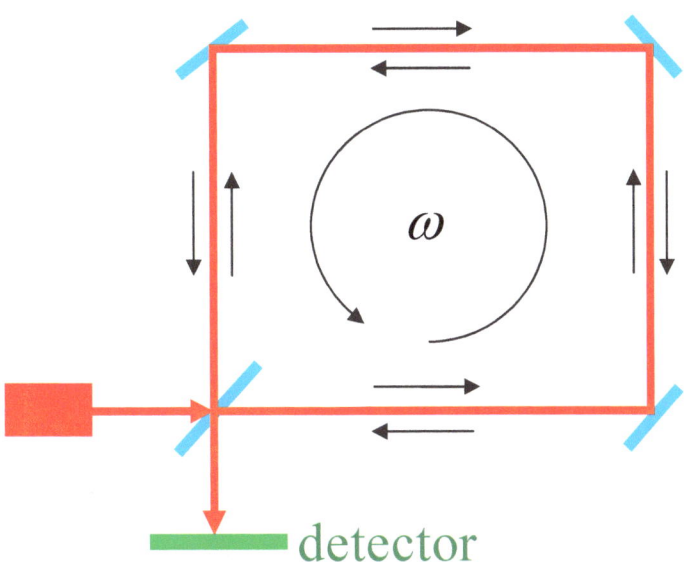

(Fig4.1-1: The basic construction of a Sagnac interferometer)

$$\Delta t = \frac{4A\omega}{c^2}$$

This actually disproves the proposition of the constancy of the speed of light though some relativists tried to explain it in some other way. However, the experiment result complies with the mathematical calculation exactly. And the calculation is based on the time difference of the two light beams traveled in opposite way. These are undoubted facts. Since there is time difference between the two coherent light beams to the same detector, this proves the two coherent light beams traveled the same path in different relative speed with respect to the detector.

4.1.2 Fiber Optic Gyroscope (FOG)

A fiber optic gyroscope (FOG) is a gyroscope based on the Sagnac effect that uses the interference of light to detect mechanical rotation. The FOG can somehow detect the Earth rotation. This seems to be contrary to our postulation that the space has no relative velocity with respect Earth rotation. However, for the Sagnac interferometer on Earth, there is another effect which will affect the photon when it's moving north-south, the Coriolis Effect.

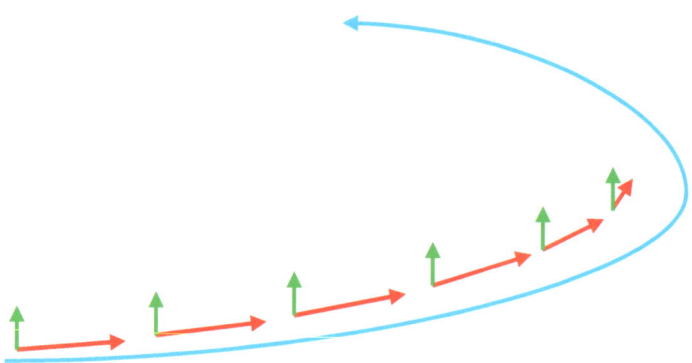

(Fig4.1-2: The Coriolis Effect on the Sagnac interferometer)

When the photon is moving in the north-south direction, the Coriolis Effect will force it to shift in east-west direction. However, the speed of light will remain constant even it is shifting transversely. Below we use a square loop to demonstrate the change of the light path.

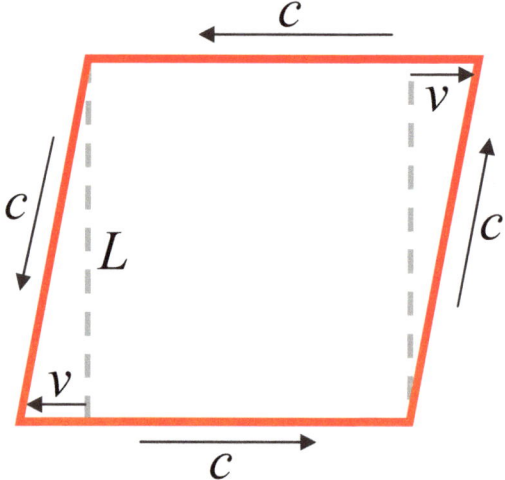

(Fig4.1-3: The Coriolis Effect at Northern Hemisphere)

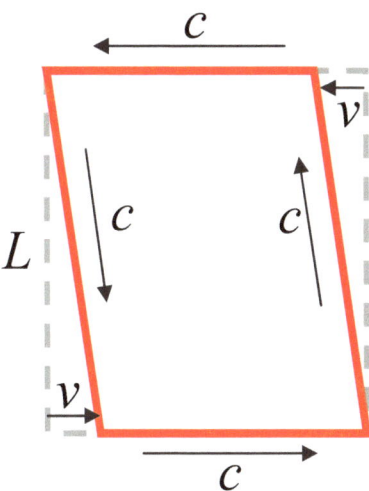

(Fig4.1-4: The Coriolis Effect at Southern Hemisphere)

Although the Coriolis Effect will not affect the light speed, however it does change the direction of the light beam. When the light beam moves for a distance L in north-south direction, it will also shift a distance $L \cdot v/c$ in east-west direction. And hence the length of the light path in east-west direction will become as:

$$L' = L \pm L \cdot \frac{v}{c} = L \cdot \left(1 \pm \frac{v}{c} \right)$$

And hence change the time for the light to travel in east-west direction. This equals to change of the light speed though even the light speed does not actually change.

$$L \cdot \left(1 \pm \frac{v}{c} \right) : L = c' \cdot t : c \cdot t \Rightarrow \left(1 \pm \frac{v}{c} \right) : 1 = c' : c$$

$$\rightarrow c' = \left(1 \pm \frac{v}{c} \right) \cdot c = c \pm v$$

In this case, the rotation of Earth detected in Northern and Southern Hemisphere will have the same magnitude but reverse direction. If Earth rotation detected on a FOG is caused by the change of light speed, the direction detected in Northern and Southern Hemisphere should be the same. If Earth rotation will change the light speed, it should be able to be detected on a sophisticated interferometer. This does not in compliance with the fact.

For the Michelson–Morley interferometers, the light beams do not travel in a circular loop but only been reflected back and forth, the shifting of light beams will not change the length of light path. And the shifting of the light projection will compensate with each other. And hence can not detect the Earth rotation.

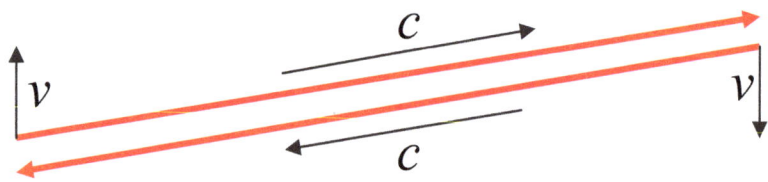

(Fig4.1-5: The Coriolis Effect on an interferometer)

4.1.3 Satellite and Earth Rotation
In the case of the geosynchronous satellites transceiving signals to the ground from an altitude of about 36,000km while Earth is rotating, the ground speed will affect the relative velocity of light since there is a relative velocity of the ground station with respect to the satellites.

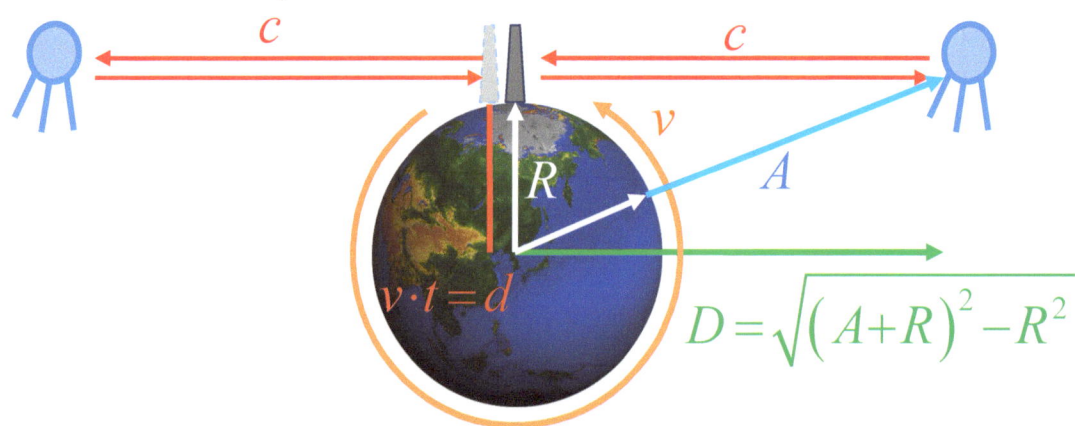

(Fig4.1-6: Satellite and Earth rotation)

In this case, the distance between the ground station and the satellite will equals to about 36,631km. And hence the electromagnetic waves will have to take about 0.25s to travel back and forth for one trip. This makes measuring of the difference of the relative velocity caused by Earth rotation possible by calculating the time difference of the signal transceiving.

Here comes a critical inconsistency to the proposition of the constancy of the speed of light. First we choose the Solar System as a rest reference, we assume the speed of light travels from the ground station to the satellite and back to the ground station is constant c, no matter it's been dragged, partially dragged or no drag. Based on this assumption, we calculate the time for the light travels in this distance.

$$2D \pm d = c \cdot t \Rightarrow 2D \pm v \cdot t = c \cdot t \qquad , \quad d = v \cdot t$$

$$\Rightarrow t = \frac{2D}{c \pm v}$$

In this case, the electromagnetic waves will have to take less or more time to travel the distance altered by the displacement of the ground station. As a result, the electromagnetic waves will then travel in different periods of time. And hence there comes different relative speeds of the light with respect to the ground station since the result $c \pm v$ is just the definition of the relative speed.

If one insists the speed of light to be constant c , $c \pm v = c$, that means there is no relative velocity between the ground station and the satellite. Not $c' \pm v = c$ because there will then have two different light speeds. In this case, the transceiving time should be independent of Earth rotation, then how will we measure the distance? One might also claim that the first assumption of the light speed to be constant c for our calculation is wrong. Therefore, the constant light speed can only be applied to either one of this example, the first assumption or the end result. This means two different values of the light speed.

4.14 Light Speed vs. Spacecraft

If we take the Solar System as a rest reference frame, for the light comes from the outer space heading towards the sun, the relative light speed with respect to the Solar System will reduce little by little gradually. The light speed with respect to the space vortex of the sun is constant c, however, the speed of the space vortex with respect to the sun complies with the equation ($v = \sqrt{GM/r}$). In this case, the relative speed of light with respect to the stationary sun u can be derived as below:

$$u = \sqrt{c^2 - v^2} = \sqrt{c^2 - \frac{GM}{r}}$$

Hence the light speed with respect to the stationary sun will decrease while it's heading towards the sun and increase while it's leaving away from the sun.

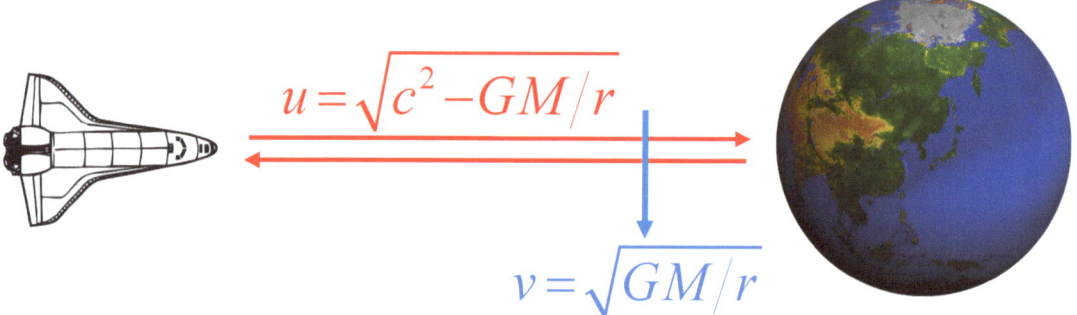

(Fig4.1-6: Signals transceiving from a spacecraft to Earth)

For a spacecraft heading towards the sun or getting away from the sun in the space, the relative velocity of the spacecraft with respect to the sun will remain constant if there is no acceleration or deceleration by any external force. In this case, if the spacecraft is transceiving the signals to the ground station on Earth while it's returning to Earth, we will perceive that the spacecraft is speeding up because the light speed is decelerating while the spacecraft remains at constant speed. On the contrary, we will perceive that the spacecraft is slowing down if the spacecraft is leaving away from Earth because the light speed is accelerating while the spacecraft remains at constant speed.

4.1.5 Conclusions

According to the examples demonstrated in this section, we can have the following conclusions:

1) As a conclusion, the speed of light with respect to its medium is constant.

2) The speed of light varies in accordance with the relative velocity of the observers with respect to the medium of light.

4.2 Gravitational Redshift and Blue-shift

Redshift and blue-shift of light waves happens when a light wave travels through a gravitational field. The light waves leaving out from the gravitational field will cause the redshift. On the contrary, the light waves entering into the gravitational field will cause the blue-shift.

When a light wave is leaving out from a gravitational field, this means it's getting away from the space vortex of the gravitational field. In a space vortex, the space is being dragged into the center of the vortex, the speed of the space flow in this direction is inversely proportional to the square of the distance. This means the speed of the space flow which pointing towards the center of the vortex will become slower with increase of the distance with respect to the vortex center.

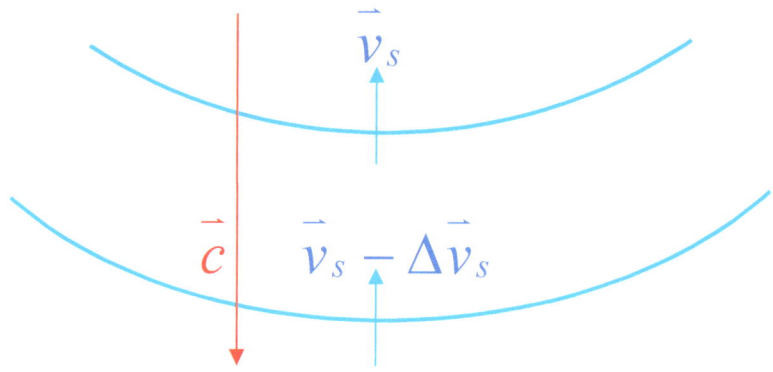

(Fig4.2-1: Light wave leaving away from a space vortex)

This can be taken as the speed of the space flow has increased for a value of $\Delta \vec{v}_s$ in the reverse direction with respect to the inner space.

Since space is the medium of light, the speed of light will be limited to the light speed in that medium and hence the light speed in the inner and outer place of the space vortex will always be the same light speed c in vacuum. In this case, the photon has to speed up to catch up with light speed when leaving away from the gravitational field.

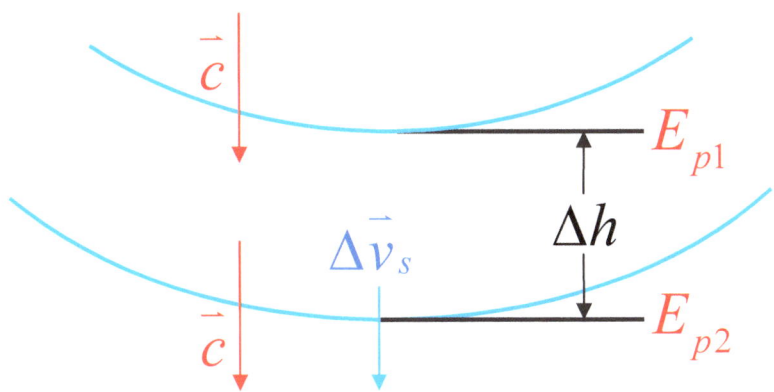

(Fig4.2-2: Light speed with respect to the speed of the space flow)

In this case, the photon has to consume some energy and transmute it into the kinetic energy to speed up. On the other hand, this can be taken as the photon is increasing the gravitational potential energy. However, the total energy of a photon should conserve with itself during traveling in the space. In this case, the energy transmutation of a photon moves for a distance of Δh in the space vortex can be described as below:

$$E_{p1} = E_{p2} + mg\Delta h$$

$$\Rightarrow E_{p1} > E_{p2}$$

$$\Rightarrow h \cdot f_{p1} > h \cdot f_{p2} \Leftrightarrow f_{p1} > f_{p2} \qquad , \because E = h \cdot f$$

$$\Rightarrow \frac{c}{\lambda_{p1}} > \frac{c}{\lambda_{p2}} \Leftrightarrow \frac{1}{\lambda_{p1}} > \frac{1}{\lambda_{p2}} \qquad , \because c = f \cdot \lambda$$

$$\Rightarrow \lambda_{p1} < \lambda_{p2}$$

In this case, the photon will lose energy but keep the light speed with respect to the flow of space and hence increases its wavelength while it's leaving away from a gravitational field. This causes the redshift to the light waves.

Comparatively, while a photon is entering into a gravitational field, it will gain some energy but keep the light speed with respect to the flow of space and hence decreases its wavelength. This causes the blue-shift to the light waves.

4.3 Construction of the 4D World

Though we can not measure time in the way we measure the distance, but we do can measure time indirectly by measuring the change of an object. That is because a 3-dimensional equipment is not capable of measuring in 4-dimensional. Anyway, the 4th time dimension is true though even we can't see it. Thus, the universe must be a 4-dimensional one actually. And our world is a 3D section in that 4D universe.

4.3.1 Construction of a 4-dimensional World

As we have defined the-future and the-past of our world as an upstream and a downstream time domain of our present world in *Chapter 1*. Though the worlds of the upstream and downstream time domains are defined as different worlds from ours, because they may not be the same as our world, however, it is also possible that some of these time domains maybe exactly the same with our world but in a little time advance and delay. This is similar to a film movie, for each frame of a video film, every single frame can be taken as an independent picture, a different 2D world. They do not have physical relationship among one another though even they are linked in a frame sequence. We see the movie plays just because we are looking at different frame pictures in a sequence.

The time domains are all linked one after one like a film. The despace holes connect between the time domains and lead the space to flow through them. This somehow makes all the time domains to have the same story. Just like all the time domains are playing the same movie with differential time difference. And makes the upstream time domains to become the same as our future, makes the downstream time domains to become the same as our past. This is similar to play several identical movies on the screens placed in series. But there is a little time difference among these movies. The movies of the upstream time domains are always precedent to those in the downstream time domains. This can be shown as the following picture:

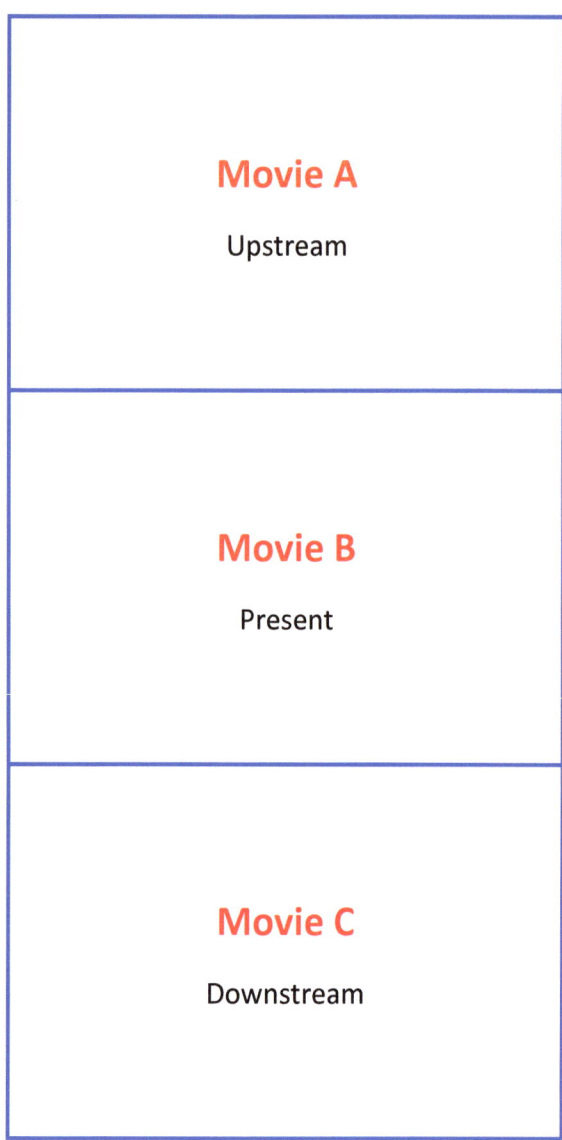

(Fig4.3-1: Identical movies played and placed together)

4.3.2 Relation between Upstream and Downstream Time Domains

According to above, how the 4-dimensional world works can be simulated as the below picture. The flowers are growing in all the time domains. They are growing in exactly the same speed. However, the flowers in the upstream worlds always grow precedent to those in the downstream worlds. When the flowers in the upstream bloom, the flowers in the downstream worlds may still in buds or maybe still be as seeds.

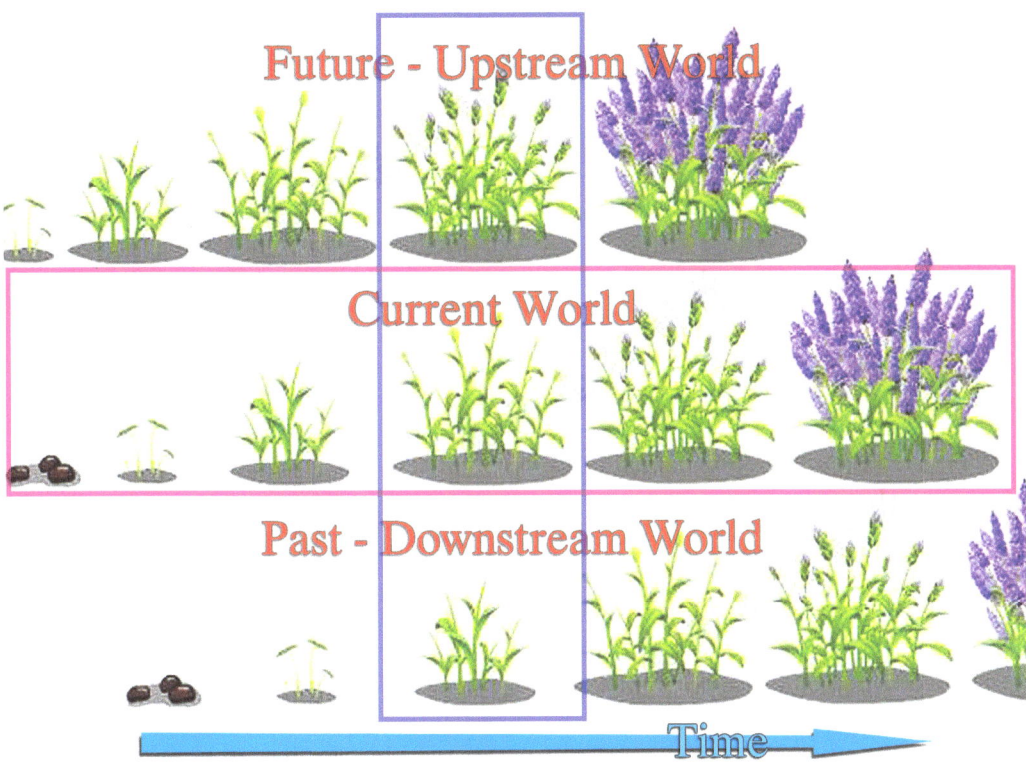

(Fig4.3-2: A simulation of a 4D world)

Since there are despace holes connecting between the upstream and down-stream time domains, the despace holes in the upstream time domain will end up in the downstream time domain and start up with another despace hole in that time domain. In this case, the two conjunction time domains will be almost exactly the same.

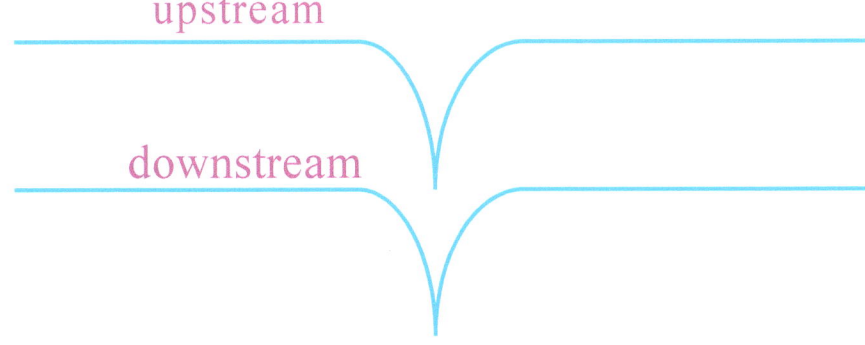

(Fig4.3-3: Despace holes between time domains)

However, since the direction of the space flow in T axis is always from the upstream to the downstream, if the despace holes in the upstream move or change 3-dimensionally, it will then affect those in the downstream. This causes the downstream world to follow after the upstream world. In this case, if there is any slight change in the upstream time domain, it will then affect all the downstream time domains to change accordingly. If the flowers wither or dry-up in the upstream world, it will then happen to those in the downstream worlds as well.

4.4 Time-Space Energy

Undoubtedly, the whole world is "moving" towards its future continuously. Everything and every corner in the world is also "moving" towards the same future in the same speed and phase. In this case, is "Time" itself some kind of energy?

We know that the Gravitation is a kind of powerful energy and it's inexhaustible though even we can not use it freely since everything will finally touches the ground. However, it's no doubt that the Gravitation is a kind of energy of the universe. According to our theory, the Gravitation is caused by the space drag. This means such energy is the power of the space flow. And this space flow flows in time dimension. This is the energy of the time-space.

4.4.1 Conservation in 4-dimensional

First of all, conservation in 3-dimensional space can be invalid sometimes, because conservation is based on time and space. That should be 4 dimensional.

We know that it requires a lot of power to drive a train to a very high speed. But when we travel on a train, we can not perceive such powerful energy. Everything on the train acts exactly the same as the train is not moving. It consumes the same amount of the energy to move on the train no matter the train is moving or not. When the train is moving in a very high speed, if we throw something out of the train, it will release a lot of energy once it touches the ground. If someone wants to throw an object onto that train from the ground, he will have to deliver energy to the object.

The universe is similar to above example. The whole universe is like the train travels in time dimension. We can not drain the energy from time dimension because we all move in the same speed in time dimension. If one can change its speed in time dimension, he will be able to drain and release energy from the 4th dimension into the 3-dimensional world. This can be described as the following:

$$E_{x,y,z} = F_{x,y,z} \cdot S_{x,y,z} = F \cdot \begin{bmatrix} S_{\Delta x} \\ S_{\Delta y} \\ S_{\Delta z} \end{bmatrix}$$, 3-dimensionally

This equals to:

$$E_{x,y,z,t} = F \cdot \begin{bmatrix} S_{\Delta x} \\ S_{\Delta y} \\ S_{\Delta z} \\ 0_{\Delta t} \end{bmatrix} = F \cdot \begin{bmatrix} S_{\Delta x} \\ S_{\Delta y} \\ S_{\Delta z} \\ S_{\Delta t=0} \end{bmatrix}$$, 4-dimensionally

We can't perceive the power of time because we are in the same speed with the whole universe in time dimension. We are on the same train.

$$E_{x,y,z,t} = F \cdot \begin{bmatrix} S_{\Delta x=0} \\ S_{\Delta y=0} \\ S_{\Delta z=0} \\ S_{\Delta t=0} \end{bmatrix} = F \cdot \begin{bmatrix} 0_x \\ 0_y \\ 0_z \\ 0_t \end{bmatrix} = F \cdot 0 = 0$$

If we can drain energy from time dimension:

$$E_{0,0,0,t} = F \cdot \begin{bmatrix} 0 \\ 0 \\ 0 \\ s_{\Delta t} \end{bmatrix} = F \cdot s_{\Delta t}$$

This will not cause the object to move in time dimension but cause the flow of space to change its flow speed in time dimension. If we drain energy E from time dimension, we make the space flow to release energy E and decelerate. If we release energy E to time dimension, the space flow will absorb energy E and accelerate.

4.5 Time Travel

I've spent all my life in searching for a way to travel in time dimension. However, after these years of researching and thinking, I realized that it is impractical to spend all the resource of mankind to send one single person to travel in time for his own good. Meanwhile, how can we make sure the time we arrive is the time we set? We can not estimate how dangerous it is to travel in time dimension.

4.5.1 Traveling

In this new era, we should have a new way of thinking. What is traveling? A man moves from one place to another one? Do we really have to be at that place to be defined as arrived?

When we position at one place, we can see the sight, hear the sound, touch things, and communicate with others where we are. However, what if we place a camera, a microphone, a speaker, some tactile sensors to simulate the feeling of touch, and some actuators to accomplish our demand at the place where we are supposed to be? And we can just sit in our office or lie on the bed to receive these information and signals. Is there any difference? We can transport the equipments rather than transport our body.

We shave everyday. We scissor nails everyday. We might lose our hands or legs and keep alive. What if someday we lose our head or our brain? Some computers have multiple processors as backups of each other. Why can't we have backup of our brain or our whole body? Then we can send our backup body to travel and send the signals back to the main body at the safe place.

4.5.2 The Echo of Time

Rather than traveling in time dimensions, I think it would be easier and more practical of finding a way to communicate between time domains. In this case, we have to find a kind of the signal which can propagate in time dimension first.

According to the Theory of Despace, the space flow caused by Gravitation in current world will affect the space flow of the other world behind our world. It may also affect the space flow of the world ahead of our world. If we can vary a certain mass of an object in a meaningful way, it will be possible to be detected as a signal of Gravitational Wave in the other world. And hence we might be able to communicate between time domains via some kind of signals like this.

There might be some other ways to communicate between time domains. Somehow, people can predict the fortune or dream what happens in the future. It seems our brain wave in the current time domain can be dragged into the down-stream time domains by the time-space flow. And hence affect the other one's (or the one's own self in different time domain) brain wave as an echo in that time domain.

How can we send messages to our future? It's simple. Write a letter and keep it at a safety place. Ask someone to remind you to read this letter on the day you appointed. Then you've just send a message to your future. Actually, how we re-call our past everyday is just as reading the message we saved in our mind in the past.

Though the worlds of the upstream and downstream time domains may be defined as different worlds from ours, however, they may also be taken as our future and past. In this case, if all the different time domains are positioned closely, the brain of each one of us in each time domain can act as an amplifier of our brain wave. And thus the thought of the one at the upstream time domain can be passed through time dimension to the downstream time domain. And hence affect the one's thought in the downstream time domains. Or in other words, our thought in the-future can affect our thought in the-past, though even they are not our real future nor the past since they are the different time domains.

As our thought being passed from the future (upstream) to the past (down-stream) and affect our thought in that time domain, and the one of ourselves in the downstream past will also bring his thought been amplified (memorized) into his own future. Thought the future of the downstream time domain is not actually our future in our time domain, however, if they are the same, if what happened in other time domains is exactly the same with what will happen in our time domain, in this case, the future of the downstream time domain can be taken as the same as our future in our time domain. This generates an echo of our thought passing back and forth in time dimension. This is the echo of time.

4.6 Time Machine

It is the dream for a lot of people to travel in time dimension. To travel in time dimension, we might need a time machine or by some other means. However, I will not discuss about how to build a time machine. I do not know how to build the time machine. Here I only discuss how the time machine might work.

4.6.1 Definition of time travel

What alleged as time travel is to transport in time dimension. This means to vary one's vector or position in time dimension.

Example A : One man spends a period of time to transport from spot A to spot B in space dimension. This man might experience different length of time than others in the world. Dose he travels in time dimension? The answer is NO.

Example B : One man freezes his own time or himself for a period of time and unfreezes later. He wakes up in the future. Dose he travels in time dimension? The answer is NO.

Within the same time domain, no matter how you slow down or speed up your time, which might means your time runs slower or faster than others, your one minute might be longer or shorter than others', however, you are still in the same time domain. You will not jump back to the past which you have experienced. You will not jump forward to the future and skip the time period which you have to experience.

Therefore, to travel in time dimension, one must be isolated from the world and shift between time domains. For example, if one wants to transport into the-future, he will have to shift into the upstream time domain with a certain length of time precedent to our world. If one wants to transport into the-past, he will have to shift into the downstream time domain with a certain length of time lag behind our world.

4.6.2 An eye hole

The only one way to isolate one thing from the world is to be surrounded by the despace 3-dimensionally as below, note that the circles would be taken as spheres in 3 dimensional space:

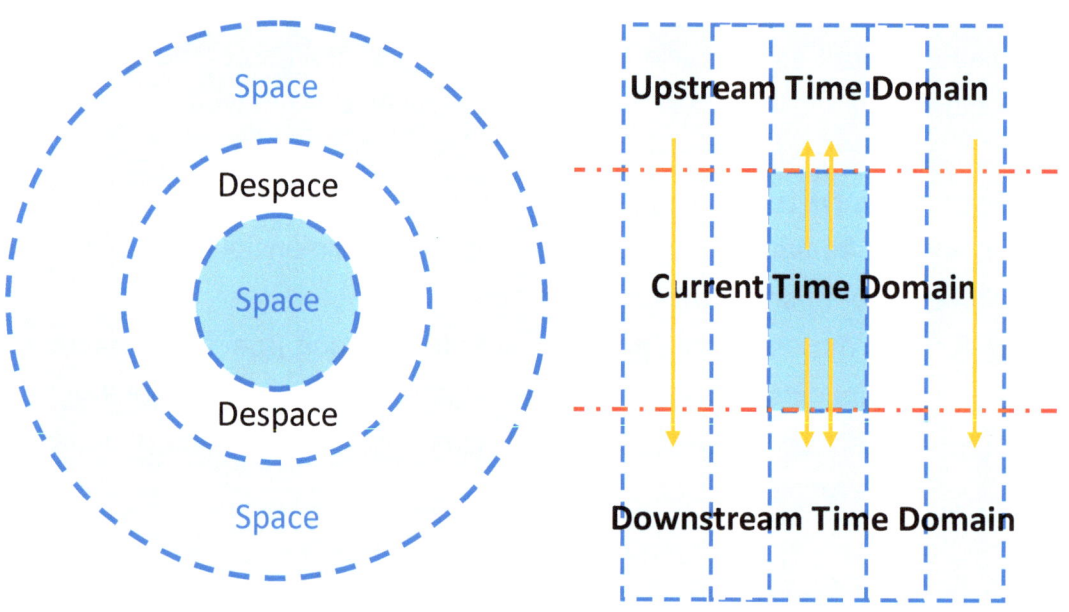

(Fig4.6-1: An eye hole)

According to figure 4.6-1, it looks like a region of the space inside a black hole. Although the space outside of the black hole will be dragged into the despace and the speed of the space flow is extremely high, however, the speed of the space flow at the center of the black hole could be very slow. It looks like the space at the center will be drained out eventually, however, the upstream time domain will also keep pouring a certain quantity of the space into the current time domain 4-dimensionally. The center space will be keep isolated if the speed of the space flow of the space being dragged into the downstream time domain and the one being poured from the upstream time domain at the center region are counterbalanced.

It looks similar to an eye of the space and hence the name an "**eye hole**". When the space in the center of the eye hole being isolated by the despace, the space flow will keep flowing in time dimension, from the upstream to the downstream time domains 4-dimensionally. If the despace gap is perfectly matched with the speed of the space flow, the space in the center of the eye hole will keep isolated. In this case, the speed of the space flow at the center of the eye hole can vary independently from the outside.

If we place one machine big enough at the center of the eye hole, and this machine can drain and release energy of the time-space in the eye hole, it will be able to control the space flow of the whole center space. We may call it as *time machine* in this article. Since the space of the whole 4D universe is flowing at a certain speed, if the speed of the space flow of the center space of the eye hole is different from the outside, it will cause the center space of the eye hole to shift up and down in time domain with respect to the outside time domains. Since the upstream time domains equal to our future and the downstream time domains equal to our past, hence shifting in time domains equals to travel in time dimension.

4.6.3 Transmitting 4-dimensionally

There is no easy task to create or eliminate a black hole, not to mention the eye hole. It's not possible to pass through the despace into the eye hole and get out. You have to destroy the black hole or to be destroyed. In other words, it's not possible to ride on the time machine to do the time travel.

However, you don't have to be actually travelled. Sometimes you would just like to experience the travel not the travelling. Just like watching a movie or playing a video game. You can have a remote controllable device in another time domain which can receive control signals from your time domain and transmit signals back to your time domain. Of course, the control signals should pass through the weakest point of the despace gap of the eye hole and then transmit into another time domain.

Normally, the space flow in time dimension is very fast, one way from the upstream to the downstream. So there is no a signal can be transmitted from the downstream time domain into the upstream time domain directly. In this case, the signals must be saved in the time machine when received. The time machine will then have to shift upwards in time dimension. The time machine will repeat the signals when it reaches the designated time domain. Normally, an upstream time domain from ours to ensure the signal can pass through the eye hole and compensate the time delay for the signal to reach the receivers. Then we can communicate between different time domains via this method and hence to simulate the time travel.

Afterword

Simple is the nature of the world. A simple thing would not be constructed from complex components. The Theory of Despace describes the world with the simplest model which no other model would be simpler than it. Our world requires only one thing to construct, the space. All the substances, energy, movement, relationships and even time, are based on it. How simple, how pure!

Matter is no matter. Particle is no particle. Wave is no wave. What we see is what to blind us. What we knew is what to deceive ourselves. What we believed is what to fool us. Only the one who breaks through those obstacles and bondages can see the truth.

What really matter in this book is not the result nor conclusion of the theory, not even the theory itself, but the innovative concept of a way of thinking and the renovating attitude toward the world.

www.ingramcontent.com/pod-product-compliance
Lightning Source LLC
Chambersburg PA
CBHW050717180526
45159CB00003B/1052